GEOLOGY UNDERFOOT
IN DEATH VALLEY AND EASTERN CALIFORNIA

Second Edition

ALLEN F. GLAZNER
ARTHUR GIBBS SYLVESTER
ROBERT P. SHARP

2022
Mountain Press Publishing Company
Missoula, Montana

GEOLOGY UNDERFOOT IS A REGISTERED TRADEMARK
OF MOUNTAIN PRESS PUBLISHING COMPANY

The Geology Underfoot series presents geology with a hands-on, get-out-of-the-car approach. A formal background in geology is not required for enjoyment.

© 2022 by Allen F. Glazner and Arthur G. Sylvester
All rights reserved
First Printing, January 2022

FRONT COVER PHOTO: *Death Valley from Zabriskie Point.*

All photographs by the authors unless otherwise credited.

Maps and figures constructed by Chelsea M. Feeney.
www.cmcfeeney.com

Library of Congress Cataloging-in-Publication Data

Names: Glazner, Allen F., author. | Sylvester, Arthur G., author. | Sharp, Robert P. (Robert Phillip), author.
Title: Geology underfoot in Death Valley and eastern California / Allen F. Glazner, Arthur Gibbs Sylvester, Robert P. Sharp.
Other titles: Geology underfoot in Death Valley and Owens Valley.
Description: Second edition. | Missoula, Montana : Mountain Press Publishing Company, 2022. | Series: Geology underfoot | Revision of: Geology underfoot in Death Valley and Owens Valley / Robert P. Sharp and Allen F. Glazner. 1997. | Includes bibliographical references and index.
 | Summary: "Authors Allen Glazner and Art Sylvester build on coauthor Bob Sharp's insights to produce this full-color illustrated guide to 33 amazing geologic sites in Death Valley and the surrounding region. —Provided by publisher.
Identifiers: LCCN 2021046210 | ISBN 9780878427079 (paperback)
Subjects: LCSH: Geology—Death Valley (Calif. and Nev.) | Geology—California—Death Valley. | Geology—California—Owens Valley. | Geology—California—Guidebooks.
Classification: LCC QE90.D35 S47 2022 | DDC 577.94/87—dc23/eng/20211026

Printed in the United States

P.O. Box 2399 • Missoula, MT 59806 • 406-728-1900
800-234-5308 • info@mtnpress.com
www.mountain-press.com

To Mary, my companion on visits to all these places and the person with whom my love for Death Valley and eastern California began.

—AFG

To the memory of Tor H. Nilsen, scholar, polymath, gourmet, friend, and geologist extraordinaire, with whom I spent many days over the years leading field trips in Death Valley and southern California.

—AGS

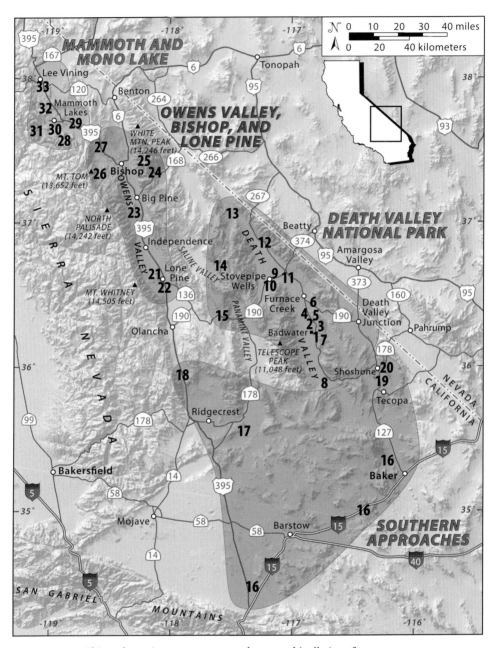

Thirty-three vignettes are grouped geographically into four areas.

CONTENTS

AN IMPORTANT NOTE TO READERS vi

PREFACE vii

Acknowledgments viii

THE SUBLIME GEOLOGY OF EASTERN CALIFORNIA 1

VIGNETTES

DEATH VALLEY

1. Badwater 9
2. Devils Golf Course and the Badwater Salt Flats 16
3. Natural Bridge Canyon 21
4. Ventifact Ridge 26
5. Artists Drive 32
6. Zabriskie Point, Gower Gulch, and Golden Canyon 37
7. Dantes View 45
8. Lake Manly 57
9. Mesquite Dunes 65
10. Mosaic Canyon 73
11. Salt Creek 83
12. Titus Canyon 89
13. Ubehebe Crater 103
14. The Racetrack 111
15. Rainbow Canyon and Father Crowley Vista Point 121

SOUTHERN APPROACHES

16. The Mojave River 131
17. Trona Pinnacles of Searles Lake 139
18. Fossil Falls 147
19. Ancient Lake Tecopa 155
20. The Resting Spring Pass Tuff 162

OWENS VALLEY, BISHOP, AND LONE PINE

21. Alabama Hills 169
22. Lone Pine Fault and the 1872 Earthquake 177
23. Big Pine Volcanic Field 187
24. Poleta Folds 195

25. Ancient Bristlecone Pine Forest   205
26. Buttermilk Boulders   217
27. Owens River Gorge   229

Mammoth Lakes and Mono Craters

28. Hilton Creek Fault at McGee Canyon   239
29. Hot Creek Geological Site   246
30. Mammoth Mountain and Long Valley Caldera   253
31. Devils Postpile   263
32. Obsidian Dome, Inyo Craters, and the Earthquake Fault   271
33. Mono Craters from Punch Bowl to Panum   281

GLOSSARY   290

LIST OF SOURCES   299

INDEX   306

ABOUT THE AUTHORS   316

GEOLOGIC CHECKLIST   318

AN IMPORTANT NOTE TO READERS

Although the publisher and authors have taken reasonable steps to ensure the accuracy and timeliness of the information contained in the book, readers are strongly encouraged to confirm details before making any travel plans. Location and direction information may change; GPS coordinates, provided for some hard-to-find sites, are approximations and should be treated as such. If you discover any out-of-date or incorrect information in the book, please let us know by sending an email to the publisher at info@mtnpress.com.

Geology Underfoot in Death Valley and Eastern California was written in the spirit of adventure, and readers are cautioned to travel at their own risk and obey all local laws. Some of the places described in this book are not open to the public and are not meant to be visited without appropriate permissions. Neither the authors nor the publisher shall be liable or responsible for any loss, injury, or damage allegedly arising from any information or suggestions contained in this book.

Please remember that collecting fossils, artifacts, plants, rocks, etc., is not allowed in Death Valley National Park and other state and national parks and monuments without a permit.

PREFACE

Eastern California is the best place in the world to study geology. Its remarkable landscape was produced by a confluence of geologic events over hundreds of millions of years, and the arid environment and high mountains have laid it all bare to see. Although this part of the state is sparsely populated, more than 30 million people live within a 250-mile radius of Lone Pine, California, near the center of the area covered in this book. Lone Pine is only a three-hour drive from Hollywood (at 3 a.m., when traffic jams are less likely), and cinematographers have featured this landscape in thousands of films, television series, and commercials, at least as far back as 1915 when Cecil B. DeMille directed *Chimmie Fadden Out West*.

Through their on-screen appearances, the soaring mountains, cliffs, boulder piles, salt flats, volcanoes, sand dunes, and other landscape features of eastern California are familiar to hundreds of millions of people, but the fascinating, action-packed stories behind how these features formed are not well known. The purpose of this book is to tell these stories in bite-size pieces—vignettes—each of which makes a good outing ranging from a few hours to a full day. Vignettes have been stripped of much of the technical detail and jargon that might go into them, and the book is aimed at people ranging from the geo-curious to professionals wishing to learn about things outside of their specialty.

There are so many stories to tell. Perhaps nowhere is evidence for natural climate change shown better than in Owens Valley, where you can stand in blazing 100-degree desert heat on a glacial deposit that was laid down perhaps only 15,000 years ago. That's not long before beer was invented. Uneroded fault scarps attest to major earthquakes. A pastoral valley east of Mammoth Lakes was produced in a cataclysmic volcanic eruption far larger than any that have occurred in human history. Many of the links between the biosphere (living things) and the lithosphere (rocky things), such as the survival and speciation of the little pupfish, are laid bare. The huge salt flat of Death Valley, and faint shorelines high on the cliffs, reveal the former existence of a 100-mile-long lake. Entire mountains composed of sedimentary rocks that have been flipped upside-down show the power of plate tectonics.

This second edition is dramatically changed from a black-and-white book first published in 1997. *Geology Underfoot in Death Valley* was the second in the Geology Underfoot series and the second collaboration between Bob Sharp and Allen Glazner. In preparing this new full-color version we had to make hard choices about what to include, and many vignettes from the first edition had to be cut or combined to make way for eleven new ones. We could easily make a companion edition with another thirty or so vignettes that focus on completely different geologic topics. Several other books, including those in our List of Sources, are must-haves for this area. We particularly recommend *Geology of Death Valley National Park* by Marli Miller and Lauren Wright; *Geologic Guidebook to the Long Valley: Mono Craters Area of Eastern California, 3rd Edition* by Steve Lipshie; and *Hiking Death Valley: Guide to its Natural Wonders and Mining Past* by Michel Digonnet.

A note about elevations. There is no definitive source for the elevations of peaks and other geographic features, and, as is evident from the tectonic history of eastern California, elevations change over time. Any inconsistencies in quoted elevations in this book reflect inconsistencies in the sources we have used.

Before you begin, a few words of caution. Much of the area covered in this book, including most of Death Valley National Park, is wild and remote. Before undertaking any excursion into the backcountry, especially to Titus Canyon or the Racetrack, do the following. (1) Check current road conditions at the Furnace Creek Visitor Center. (2) Bring enough food and water to last your intended time and a few more days. (3) Expect to get a flat tire; make sure that you have an adequate spare tire and can remove the lug nuts on each wheel (this last item comes from one of your authors during one of his flats on the way to the Racetrack). Finally, note that mobile phone service is lacking in much of eastern California. You cannot call for a tow truck from many of these places, and mobile maps are unavailable without preplanning.

Grab this book, some food, a good road map, and lots of water, and follow these vignettes. You will see the landscape in new ways.

Acknowledgments

Rivers do not arise with the first rainfall, and rarely does a book arise from its first draft; thus, we must acknowledge and thank the many people who had a hand in helping us produce this second edition.

The first two authors were introduced to the spectacular geology of eastern California by the third author, Bob Sharp, and by Clem Nelson, Bob Webb, Bob Norris, Steve Lipshie, Roy Bailey, Dave Hill, and Wes Hildreth, among many others. Kurt Frankel (1978–2011), beginning with his undergraduate enthusiasm for the Racetrack, nucleated a number of magnificent field trips that are represented in these pages and led to his career in geology and his contributions to our knowledge of faulting in Death Valley. The chapters herein are based on a great deal of field work over many years, and we were accompanied and helped on these excursions by excellent field companions including Cecil Patrick, Steve Lipshie, Tim Forsell, Scott Hetzler, Bob Harrington, Jeff Glazner, Hunter Gallant, John Bartley, Drew Coleman, Tor Nilsen, Ryan Taylor, Clementine, Scooter, and many dozens of students.

Vignettes were field-checked and critiqued by experts Jeff Knott, Darrel Cowan, Bob Harrington, Marli Miller, Doug Walker, Kurtis Burmeister, Hildy Schwartz, Anne Egger, Stan Finney, Jorge Vazquez, Wes Hildreth, Bruce Lander, Nathan Niemi, Connie Millar, Phil Pister, Dick Norris, Jim Norris, Steve Lipshie, Kurt Fausch, Cecil Patrick, and Rodney Thompson.

The White Mountain Research Center provided a base for much of this work, and its superb staff, including Denise Waterbury, Jeremiah Eanes, Steven Devanzo, Tim Forsell, and Daniel Pritchett, went well beyond their job descriptions to help. We also gratefully acknowledge Darrel Cowan's hospitality and Doug Walker's 1992 Ford Ranger.

We have again had the good fortune to work with Jenn Carey (editor), Chelsea Feeney (illustrator), and Jeannie Painter (layout), who magnificently melded our prose and illustrations into this book.

Kurt Frankel studying one of the young fault scarps cutting the alluvial fan just south of Badwater in 1999.

EON	ERA	PERIOD	EPOCH	AGE (mya)*	EASTERN CALIFORNIA EVENTS IN GEOLOGIC TIME	FORMATION NAME
PHANEROZOIC	CENOZOIC	QUATERNARY	HOLOCENE		1872 earthquake Inyo/Mono Craters Ubehebe Crater Mammoth Mountain Devils Postpile	
			PLEISTOCENE	0.01	Long Valley resurgent domes — Sierran glaciations and pluvial lakes, including Lake Manly and Trona Pinnacles eruption of Bishop Tuff	
		NEOGENE	PLIOCENE	2.6	Glass Mountain volcanism opening of Panamint Valley	Furnace Creek Formation
		(TERTIARY)	MIOCENE	5.3	opening of Death Valley	Artist Drive Formation
				23		ash-flow tuffs
		PALEOGENE	OLIGOCENE	33.9	rocks of these ages are not found in Death Valley and eastern California	Titus Canyon Formation
			EOCENE	56		
			PALEOCENE	66		
	MESOZOIC	CRETACEOUS			intrusion of Sierra Nevada batholith, thrust faulting; ~200-65 mya	Alabama Hills granite
		JURASSIC		145		Independence dike swarm
		TRIASSIC		201	Alabama Hills volcanism	
	PALEOZOIC	PERMIAN		252		
		PENNSYLVANIAN		299		
		MISSISSIPPIAN		323		
		DEVONIAN		359	continental margin sedimentation ~800-200 mya	
		SILURIAN		419		
		ORDOVICIAN		444		
		CAMBRIAN		485		Bonanza King Carrara Zabriskie Wood Canyon Harkless Poleta Campito
				541		
PROTEROZOIC				2,500	formation of metamorphic and igneous rocks of the Black Mountains; ~2,000-1,100 mya	Stirling Johnnie Deep Spring Reed Noonday
ARCHEAN				4,000	rocks of these ages are not found in Death Valley and eastern California	
HADEAN					formation of the Earth; 4,600 mya	

Annotations spanning the Cenozoic: intermittent Lake Tecopa ended ~200,000 years ago; Basin and Range stretching ~800mya-present; subduction of oceanic plates.

*mya = millions of years ago

THE SUBLIME GEOLOGY OF EASTERN CALIFORNIA

Eastern California is arguably the most geologically diverse place on Earth. It has nearly everything a geologist could want, including young volcanoes, frequent earthquakes, recent glaciation, a record of the rapid diversification of life 540 million years ago, glacial deposits from a time when the entire planet was iced over, sand dunes, hot springs, slot canyons, dried beds of former desert lakes, strange tufa towers, and thick volcanic ash beds from distant supereruptions. With its sparsely vegetated exposures, good weather, and excellent roads, it is no surprise that people come to eastern California from all over the world to study geology. Along the way, they'll be treated to extreme topography with the highest peak in the lower 48 states rising above the lowest place in North America.

If you have studied a shaded relief map of the western United States, perhaps you, like American geologist Clarence Dutton in an 1886 US Geological Survey report, have noticed that the mountains of Nevada and surrounding states resemble "an army of caterpillars crawling northward out of Mexico." These caterpillars are mountain ranges, the spaces in between them are valleys, and together they form the vast Basin and Range province of western North America.

The Earth's crust underlying the Basin and Range has been tectonically stretching over the past several tens of millions of years and has moved the Sierra Nevada about 150 miles to the west-northwest relative to the Colorado Plateau. Nowhere is the evidence for stretching better displayed than in Death Valley and the area between it and the Sierra Nevada, including Panamint, Saline, and Owens Valleys.

Crustal extension occurs in several ways. In the Basin and Range, normal faulting is responsible for most of the topography by a process akin to how upright books on a bookshelf may tilt and spread out if bookends are moved apart. The surfaces among the books are the fault surfaces along which slip occurs. The blocks, and any layers in them, are tilted during this process, and the peaks of the tilted books are the ranges of the Basin and Range. In Death Valley this block tilting is expressed especially well by tilted layers of sedimentary rocks. As you

The area covered in this book lies in the western part of the Basin and Range province.

(Left) Colored bands represent initially horizontal layers of sedimentary rocks between bookends representing the Sierra Nevada on the left and the Colorado Plateau on the right. The vertical surfaces among the books are potential faults. (Right) As the Sierra Nevada bookend moves westward relative to the Colorado Plateau bookend, the books slip on the tabletop, which is the master fault, and are tilted so that their sedimentary layers tilt uniformly eastward. The surfaces between the books are normal faults, and the zigzag topography of the top surface represents basin-and-range topography. If the books were less rigid, the normal faults would be curved surfaces that merge into the tabletop surface.

tour Death Valley, notice how most of the rock layers are inclined to the east as if they were a long succession of tilted books on an enormous bookshelf.

Triangular gaps open where the bases of the rigid books meet the tabletop. In the real world, the rocks are broken by myriad small faults so that the steep normal faults curve as they pass downward into the low-angle (relatively flat-lying) fault, known as a detachment fault, represented in our bookshelf example by the surface under the books. Detachment faults were once thought to be mechanically impossible geologic structures, but scientists found them here in Death Valley and

Tilted rock layers on the north side of Furnace Creek Wash, looking north, beneath a gazebo known as the Tea House (upper right). Young layers such as these are tilted generally eastward over a large area around Furnace Creek and CA 190 southeast of there.

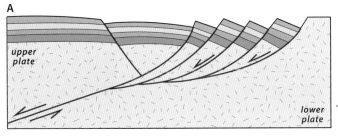
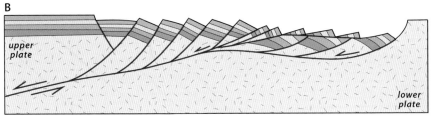

Schematic diagram of normal faults merging downward in a master detachment fault. All these faults accommodate stretching of the crust.

elsewhere in the Basin and Range. You may lay your hands upon one in Natural Bridge Canyon, the subject of Vignette 3.

Volcanism, another response to crustal extension, is dramatically expressed in the Mammoth Lakes and Mono Craters regions, where virtually every hill is a young volcano. Volcanism is also present at several other sites in this book, including Ubehebe Crater in Death Valley and the Coso and Big Pine volcanic fields in Owens Valley. Extension and volcanism commonly go hand in hand worldwide, as they do here. Does extension localize volcanism by breaking the crust and allowing magma to reach the surface, or does volcanism facilitate extension by softening the crust and allowing it to stretch? Eastern California is a prime place to study this chicken-or-egg problem.

FAULTS AND FOLDS

The crust responds to stress by breaking (faulting) and folding, and the style of these structures reflects the stress that formed them. A fault is a rock fracture along which the rocks on either side have slipped relative to one another. Faults are named according to how the blocks of rocks on either side of the fault are displaced. Strike-slip faults are those whose blocks slip horizontally relative to one another along a generally

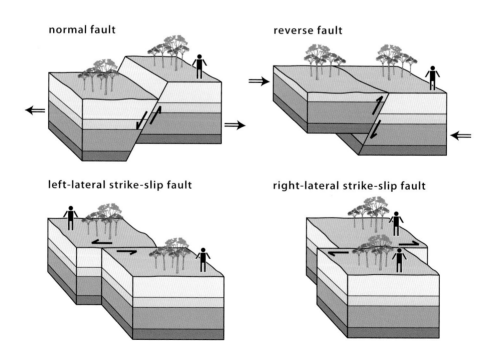

Geometry of various fault types. For the strike-slip faults (lower row), each person sees the other move left or right, respectively. Actual faults typically involve combinations of movements.

steep fault surface; they are termed right-slip or left-slip faults according to which direction the blocks slip relative to one another. Horizontal slip may be as much as hundreds of miles.

Another class of faults, dip-slip faults, involves vertical displacement. If the blocks are being pulled apart, as in the Basin and Range, then normal faults are produced—"normal" because in the place they were first defined, the coal beds of England, they were the most common type. This is also true in the Basin and Range. If the blocks are pushed together, reverse faults result. Particularly flat-lying reverse faults are known as thrust faults, and they are responsible for much of the intense deformation seen in older rock sequences in eastern California mountains, as in Titus Canyon (Vignette 12).

Rocks deformed by tectonic forces often respond by folding, which also produces tilted rocks. Folding is generally, but not necessarily, a response to shortening of the rocks. Although any type of rock can be folded under the right conditions of temperature and pressure, layered sedimentary rocks display folding particularly well. A-shaped folds are anticlines, U-shaped folds are synclines, and folds that have been deformed enough that the two sides of the fold are nearly parallel are isoclinal folds.

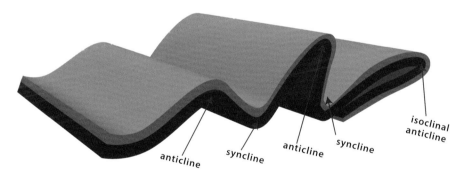

Schematic representation of two upright anticlinal and synclinal folds and one isoclinal fold.

GEOLOGIC TIME

Before we begin our study of eastern California geology, it is important to appreciate the immensity of geologic time, a concept that even experienced geologists and astrophysicists have trouble coming to grips with. The age of the solar system has been determined by several different means, giving a consistent value of 4.6 billion (4,600 million) years. A simple walk can illustrate the enormity of geologic time, and we encourage you to read this as you take a walk of about one-quarter mile, representing the 4.6 billion years of Earth's history. If each step represents 10 million years, then it takes 460 steps to get from 4.6 billion years ago, when the solar system and Earth formed, to the present.

Let's begin. Your starting point represents the formation of the solar system and Earth at 4.6 billion years. Walk 80 steps to 3.8 billion years ago, the age of the oldest known rocks discovered so far on our planet. Walk another 260 steps, a vast time of 2.6 billion years, to reach 1.2 billion years ago and the first known jellyfish-like animals. Another 66 steps bring you to the Cambrian Period, 540 million years ago, when life on Earth rapidly diversified. Twenty-one more steps bring you to 330 million years ago when the first reptiles emerged. Another 27 steps get you to 60 million years ago and the first primates. Take 6 more steps, 460 total, to reach the present day. Draw or imagine a pencil line on the ground at the end of your journey. The thickness of that pencil line corresponds to the length of recorded human history, about 5,000 years.

The 4.6 billion years of geological history are divided into five major divisions. From oldest to youngest they are Hadean (4.6 billion to 3.8 billion years ago), Archean (3.8 billion to 2.6 billion years ago), Proterozoic (2.6 billion to 541 million years ago), Paleozoic (541 to 252 million years ago), Mesozoic (252 to 66 million years ago), and Cenozoic (66 million years ago to the present). In these vignettes we are concerned principally with rocks, events, and features of Paleozoic and Cenozoic times, even though California contains rocks as old as 1.7 billion years.

Geologic time is measured in many ways. The most reliable and widely applicable is by measuring the disintegration products of radioactive elements in minerals and rocks. The most useful and reliable clocks involve the slow disintegrations of potassium to argon and uranium to lead. All igneous rocks contain traces of these elements, and by careful, precise analyses, it is possible to read these geological clocks in many rocks.

A Very Brief Geologic History of Eastern California

The following greatly simplified geologic history applies to much of Death Valley and eastern California:

1. The area was in the middle of a supercontinent, Rodinia, about 1,000 million years ago, more akin to modern-day Kansas than California.

2. Rifting, or the breaking apart, of that continent about 800 million years ago produced a new ocean basin and placed present-day California at the continental margin. Rifting of the supercontinent Pangaea around 200 million years ago, to form the Atlantic Ocean, is a more recent example of the same process.

3. A thick sequence of marine sedimentary rocks was deposited on the new continental margin over the next several hundred million years as the margin subsided. These rocks are widespread in eastern California.

4. About 230 million years ago the oceanic plate off the west coast began to descend beneath the continent. This process of subduction produced copious magmas, some of which found their way to the surface to erupt as volcanic rocks. Other batches of magma cooled and crystallized underground and produced the vast Sierra Nevada batholith, a conglomeration of thousands of granitic magma bodies. Subduction also caused dramatic east-west shortening as rocks were transported eastward along thrust faults.

5. Over the last 40 million years or so, this stack of rocks has been eroded by rivers, buried by volcanic rocks, and stretched along normal faults, producing Basin and Range topography.

Today, California straddles the junction between the Pacific and North American Plates, two of the largest and most active tectonic plates. From northern California south to the Gulf of California, the plate boundary is nominally the San Andreas fault, where the great Pacific Plate slides to the northwest against North America, grinding, shaking, and shuddering in great earthquakes.

About 75 percent of that displacement occurs along the San Andreas fault, but the Pacific Plate is also pulling away from the North American Plate, resulting in stretching that produced the Basin and Range province. The remaining 25 percent of the displacement is distributed along faults throughout the Basin and Range.

Volcanism has shaped much of the landscape of the eastern Sierra Nevada and Basin and Range for more than 40 million years. Volcanoes produce both lava flows and tephra, a convenient term for magma that is fragmented and blown out of a volcano during an eruption. The term encompasses material ranging from fine volcanic ash to volcanic bombs the size of an automobile. Volcanoes in the Mammoth Lakes area, one center of current activity, erupted as recently as 400 years ago (Vignettes 32 and 33), and Long Valley was the site of a colossal eruption and caldera collapse 767,000 years ago (Vignettes 27 and 30). Earthquakes and enigmatic emission of carbon dioxide around Mammoth Lakes since 1980 seem to indicate that the area is still volcanically active. Young volcanoes also abound in the Coso Range north of Ridgecrest (Vignette 18), and other fields of young volcanoes dot the landscape south of Big Pine (Vignette 23) and at Ubehebe Crater in northern Death Valley (Vignette 13).

PLUVIAL LAKES AND GLACIAL EVENTS

For the past 3 million years or so, the Earth's climate has been oscillating between relatively warm and dry (as it is now) to cooler and wetter over cycles of around 40,000 to 100,000 years. Glaciers formed during those cooler and wetter cycles, known as pluvial periods, and advanced in the mountains. Many of the dry lakes that dot the western deserts were full of water. Some overflowed and were fresh; others had no outlets and were salty. As the climate shifted from a glacial cycle to a warmer one, all the lakes dropped to where they were no longer overflowing, and most disappeared, leaving behind salty mud flats known as playas. These former lakes are known as pluvial lakes. Geologists refer to pluvial lakes with "Lake" first to distinguish them from modern playas; e.g., Lake Searles is the pluvial ancestor of modern Searles (dry) Lake.

Several prominent episodes of glaciation have occurred in the past few million years, but conspicuous evidence exists for only the last two. Little is known of earlier ones because later ones erased most of the evidence for them. In eastern California the last two glacial episodes are called the Tioga glaciation, which peaked about 20,000 years ago, and the Tahoe glaciation, which peaked about 140,000 years ago. The Tioga glaciation is commonly referred to as the Last Glacial Maximum. Glacial moraines, ridges and piles of bouldery debris left behind when glaciers melted, are abundant in the canyons on the east side of the Sierra Nevada. These telltale landforms are easily seen along Bishop Creek (Vignette 26), Pine Creek (Vignette 27), and McGee Creek (Vignette 28).

In eastern California, geologic processes are alive and well. Flash floods and debris flows triggered by rainstorms will continue to move sediment from mountains to valleys. Strong earthquakes will produce surface ruptures and slide the western part of California northwestward. Stones will slide on the Racetrack. Sand will blow this way or that. And who knows, a volcano might erupt.

VIGNETTE 1

BADWATER

How Low Can You Go?

You made it. You reached the lowest point that you can drive to and are not far from the lowest point, period, in North America. The 1986 US Geological Survey topographic map of this part of the valley had an informally measured elevation of -279 feet a few hundred yards west of the parking area, and a more formal measurement of -282 feet, the oft-quoted lowest point, about 3 miles to the west-northwest. Later maps leave these points off, and once you walk out onto the untrampled salt flat here or at Devils Golf Course (Vignette 2) you'll see why—the surface is rough and ever-changing as the salt crust dissolves, forms again, and heaves up.

Badwater isn't the lowest place in the world. In the eastern hemisphere the shore of the Dead Sea in Israel and Jordan is considerably lower at -1,371 feet, and in South America the valley bottoms of sediment-filled fjords are more than 300 feet below sea level. No matter, Badwater is really low.

The pool of water at the Badwater parking lot is fed by a spring that comes to the surface along a fault that follows the base of the mountain front. The water, highly enriched in salt, is certainly bad but probably isn't poisonous. Even though it is saltier than seawater, pickleweed grows around a pond full of insects and little black Badwater snails.

The salt flat (Vignette 2), also known as a salt pan, is principally table salt, or halite (NaCl), which forms a layer several feet thick on top of salty muds. The salt pan has been crushed flat under countless shoes near Badwater, but the farther you walk the more untrammeled it is, and the salt will crunch under your feet and stick to your shoes. You may walk far enough to where the salt is cracked by expansion during recrystallization. Look for salt pinnacles that grow along the cracks where brine is carried up from the water table by capillary action, evaporates, and precipitates salt on the pinnacles. About 2 miles from the parking lot, walking becomes more difficult because you'll encounter great slabs of salt that have buckled up with curled edges. Here and there you may sink up to your shins in sticky mud. What could be more fun?

Infrequent flash floods and debris flows roar down Death Valley's gulches, washes, and canyons carrying their load of water, mud, sand,

VIGNETTE 1: BADWATER

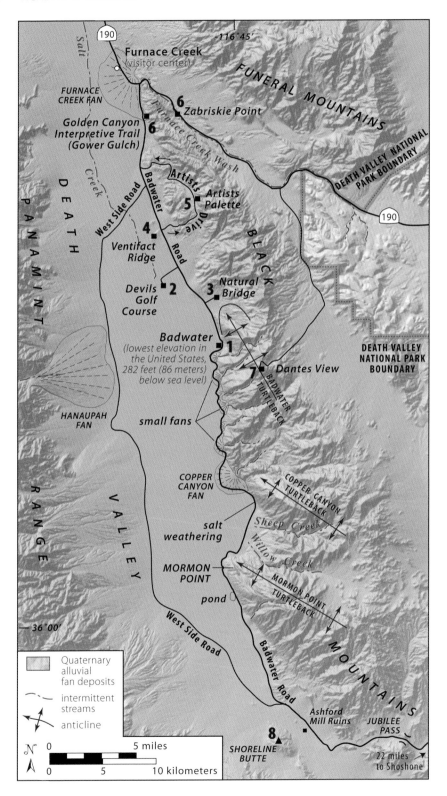

GETTING THERE

Drive south 17 miles from Furnace Creek down the Badwater Road to the parking area. From the south, Badwater is 30 miles north of Ashford Mill and 55 miles from Shoshone via CA 178. Restrooms are in the parking area.

Salt surrounds the spring-fed pool at aptly named Badwater, which lies at the steep base of the Black Mountains. Arrows point out cemented gravels that were shoreline deposits of pluvial Lake Manly.

The walk out onto the Badwater salt pan is popular, but factor in temperature when you decide how far to go. The Panamint Range, with huge alluvial fans along its eastern flank, is in the background.

gravel, cobbles, and boulders. Upon reaching the mouth of a canyon, a flood loses velocity, so it no longer has the energy to carry a heavy sediment load. The result is that a flood spreads its alluvial load out onto the valley floor in a fan-shaped deposit called an alluvial fan.

Aerial view of an alluvial fan emerging from a deep, narrow, canyon just south of Badwater. Arrow points at mysterious troughs at the toe of the fan (see discussion on page 15).

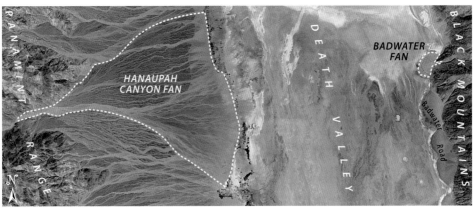

Plan view and profile view comparing sizes of Death Valley's west (left) side and east side alluvial fans, as exemplified by Hanaupah and Badwater fans.

Alluvial fans in central Death Valley are noteworthy in that they are comparatively small along the steep eastern front of the Black Mountains, whereas they are absolutely huge across the valley on the long western slope of the Panamint Range. The same fan asymmetry exists in Panamint and Owens Valleys—small fans along their east sides and much larger fans coming off their long, sloping west sides, so large that they coalesce into long alluvial aprons called bajadas.

The asymmetry of the fans in these valleys is a consequence of the asymmetry of the basin that is created by block tilting. Not only is the mountain front abruptly steep and precipitous along the east-side fault, but the valley is also deepest there, and that is where the thickest sequence of sedimentary fill is located. The basin subsides faster on the east side, so fans issuing from the steep mountain front stack upon another instead of building out into the basin. The long, gradual slope of the tilted block (Panamint Range) across the valley is gently inclined, and sediment must travel a long way east to reach the lowest spot in the basin.

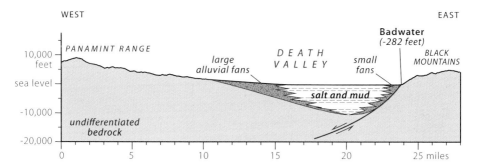

Schematic cross section across Death Valley illustrating how Panamint Range fans are built across the long, gentle eastern slope of the range. Black Mountains fans dump into the deeper eastern part of the basin, forming a stack of alluvial fans that are buried almost as fast as they build up, because slip on the fault on the east side of the range continually creates space for alluvial fans to fill.

Once you have walked a ways out onto the salt pan, pause for a few moments and ponder the asymmetry of the central part of Death Valley. Look first toward the Panamint Range and marvel at its enormous alluvial fans. They are many miles long and wide, and merge into a huge bajada. Then turn around to compare them with the smaller fans along the west base of the Black Mountains. Drilling and geophysical studies have found that the sediments are 10,000 feet thick here on the east side of the valley. In other words, the rocks 10,000 feet beneath your feet are equivalent to those at Dantes View, 6,000 feet almost directly above the Badwater salt pool. That means that vertical displacement along faults at the base of the Black Mountains adds up to about 3

The Badwater salt flat hugs the east side of Death Valley in this aerial view to the south. Badwater Road squeezes between the salt flat and the gullied, planar mountain front, which is a major fault. Badwater lies at the base of the dark cliff where the road makes a westward bend to skirt the first of five beautiful little alluvial fans.

miles. The sedimentary fill thins to the west, such that the subsurface basin is wedge-shaped. The asymmetrical shapes of basins in the Basin and Range are discussed further in Vignette 7.

After your walk out onto the playa and return to the parking lot, you may decide it is a good thing that Badwater lacks an official thermometer, so you won't know just how hot it was when you were out on the shimmering salt pan. Furnace Creek reached 130 degrees Fahrenheit on August 16, 2020, and again on July 9, 2021. Based on elevation alone, Badwater should be about 1 degree hotter, but with reflection off the salt pan and the steep mountain front it could be even hotter. This is not a particularly fun place to be on a hot July afternoon, although participants in the Badwater 135 ultramarathon, a 135-mile mid-July race from Badwater to halfway up Mt. Whitney, seem to think so. They do, however, begin their journey in the evening and run through the night to avoid the worst heat and sun.

Now face the mountain front behind the parking lot and look for the sea-level sign, 280 feet up. Along the face of the mountain near the level of the sign, you'll see faint horizontal lines that are remnants of at least one of Lake Manly's cemented shoreline gravels still adhering to the rock surface. Lake Manly, a pluvial water body, filled Death Valley on several occasions during the wetter, cooler climate of the Pleistocene

Epoch. Out of view are even higher and fainter gravel layers and Dantes View, which is more than 1 mile up.

The mountain front north of Badwater is the flank of the Badwater turtleback, a major fault surface responsible for uplift of the Black Mountains and much more. Turtlebacks and associated faults are covered in Vignette 7. A good place to put your hands on this fault surface is in Natural Bridge Canyon (Vignette 3).

Before leaving Badwater, look at the profile of the small alluvial fan that the road bends around on its way south. It is cut by two fault scarps, steep faults related to the main turtleback fault. This fan, coarser and steeper than most, was built by debris flows disgorged from the small canyon at its head. At its toe it overruns the salt pan.

The fan at Badwater is unusual in having several odd trenches near its northwestern margin (see photo on page 12). Drive 0.5 mile south from the Badwater parking area to a large pullout on the left side. A small trench is there, but better ones are next to the road on the salt pan side. These features appear to be bounded by faults that responded to northwest-southeast stretching, producing a miniature version of the Basin and Range. They die out quickly and do not align with range-bounding faults. One way these troughs may have formed assumes that the fan had built out over the salt and mud on the valley floor. If so, then shaking during large earthquakes could have sent the toe of the fan sliding out over the squishy sediments. This downslope sliding would cause the mass of fan sediments to be stretched, producing little normal faults.

Two fault scarps (the eroding steps near the base of the rocky mountain front) cut the steep, coarse alluvial fan immediately south of Badwater.

VIGNETTE 2

DEVILS GOLF COURSE AND THE BADWATER SALT FLATS
Not Many Birdies Here

The great Badwater salt flat, or salt pan, is one of Death Valley's best-known features. From a distance it looks uniformly flat, smooth, and pristine white. Close-up inspection shows, however, that it is anything but uniform. Small-scale surface features, such as the knobs and hollows at Devils Golf Course, are rugged and difficult to walk upon and would make a truly infernal golf course. Areas not recently flooded develop polygonal cracks that evolve into concave-up salt saucers. The surficial solid salt layer, 1 to 6 feet thick, lies upon salty mud. Where silt and clay adulterate the surface salt, as along the pan's edge or in stream channels and floodplains, the surface has an uneven puckered crust, like a rich cookie or pie. West Side Road crosses Salt Creek, an intermittent stream course, and many shallower stream channels border higher ground. Gentle folds and faults of small displacement deform the pan.

The 1.4-mile drive down Salt Pool Road takes you to Devils Golf Course, but you will have to search hard to find any salt pools because they are ephemeral and soon crust over and disappear. The 1938 Works Progress Administration guidebook to Death Valley said, "The salt forms so rapidly that the pools must be blasted each year"—a procedure that is not in keeping with current park policy. Instead, Devils Golf Course is a great expanse of rough and pitted salt mounds at the top of a sequence of interlayered salt and mud beds at least 1,500 feet thick. Groundwater constantly seeps into these layers after passing through the alluvial fans shed from the high mountains on both sides of the valley. As

GETTING THERE

The turnoff for Devils Golf Course from Badwater Road is 11.0 miles south of its junction with CA 190 and 2.4 miles south of the entrance to Artists Drive. The maintained gravel road (Salt Pool Road) dead-ends after 1.4 miles at a vast salt badlands. A less-visited area is easily accessed along a stretch of the West Side Road 3 to 4 miles east of Badwater Road. The West Side Road turnoff is 6.0 miles south of CA 190, between the two ends of Artists Drive. See map on page 10.

groundwater rises to the surface via capillary action, bringing dissolved salts along, it evaporates and the salt crystallizes, causing the salt crust to buckle and crack along fractures and to form delicate salt pinnacles. Wind and rain erode the pinnacles into points and edges that are sharp enough to draw blood if you fall and pour salt in the wound for good measure. If you are fortunate enough to visit on a calm, windless day, find a quiet place away from people and vehicles. Listen carefully, and you may hear the salt snap, crackle, and pop during this process of crack formation and salt crystallization.

Areas of the salt pan that are regularly flooded by streams or by precipitation runoff tend to be smooth. Some salt dissolves with each flooding

Rain erodes salt into wicked-looking, sharp-edged shapes, and evaporation produces tiny hairlike filaments that project from the surface. Fingertip for scale.

The surface of the Devils Golf Course consists of a rough jumble of closely spaced blocks and mounds about 2 feet high. Their scalloped, sharp edges would be deadly to fall upon were they carved in a harder substance. This view looks southwest across Devils Golf Course to the distant Panamint Range and the huge bajada along the base of the range.

SALT WEDGING

Anyone who lives in a cold climate or visits high mountains knows the damage that frost wedging can do to streets, sidewalks, and bedrock. Frost wedging occurs in rocks when water seeps into cracks, freezes, and expands, gradually wedging the cracks apart. In southern Death Valley a similar phenomenon, salt wedging, breaks down rocks. It happens near the edge of the salt pan after rain when salty water seeps into cracks, evaporates, and deposits tiny salt crystals that wedge the cracks open. Over time the rocks fall apart. This process is far more efficient in layered rocks such as schist than in hard rocks with few cracks, such as many volcanic rocks. Salt weathering is especially pronounced around Mormon Point. To see salt weathering wreaking havoc, drive south from Badwater and around the huge Copper Canyon fan. About 0.8 mile after the road rejoins the mountain front, it traverses a low double-walled roadcut through the toe of a debris flow (36.1024, -117.7322). Park, walk upon the debris flow surface, and see rocks suffering slow but incessant destruction.

Salt wedging has reduced a piece of black schist above the knife to a series of crumbly thin plates, whereas the more competent gray volcanic rock to the left of the knife is much more intact. All will eventually be reduced to sand by salt wedging.

These remarkably intricate patterns of salt, ever-changing as precipitation dissolves and reprecipitates it, form when wind blows salt-saturated water from ephemeral lakes over mudflats. The water evaporates, leaving white salt deposits behind. The width of this aerial view, taken in March 2006 about 4 miles north of Furnace Creek, is about 1 mile.

and precipitates again as a smoothing veneer when the water evaporates. Strong winds can blow these shallow salt pools onto muddy ground, producing exquisite, scalloped patterns when viewed from the air or from the next-best place, Dantes View (Vignette 7). Extensive flooding of the pan's surface, as happens in wet years such as 1969, 1995, and 2005, can completely erase a field of saucers. In such years a lake 1 to 3 feet deep, suitable for kayaks or canoes, covers Badwater Basin.

Salt Saucers

For a more peaceful study of salt along a stretch of ephemeral Salt Creek, drive about 2 miles southwestward along West Side Road from Badwater Road (the well-marked turn is 1.1 miles south of the Artists Drive exit). The first 1.5 miles descends an alluvial fan surface and then hits brown mud-covered salt. Four-tenths of a mile farther, the mud gives way to bright white salt carved into polygons a few feet across. This ribbon of salt runs down the axis of the valley and is the channel of Salt Creek—dry most of the time but wet enough to move the mud and freshen the salt surface. You will find plenty of interesting salt features, such as polygonal salt saucers, and, if you are lucky, clear, open pools of brine.

Looking south along the dry channel of Salt Creek from West Side Road. Salt precipitates in cracks along the edges of salt saucers, causing them to grow laterally until they jam into one another. Their edges heave up and form rugged ridges wherein continued salt precipitation causes them to grow and buckle. The average width of the polygons is about 4 feet.

Salt crystallizes along the edges of a brine pool where the West Side Road crosses Salt Creek (36.3438, -116.8648). The view is eastward toward the Black Mountains. Artists Drive is in the brightest section of rock in the center of the image.

VIGNETTE 3

NATURAL BRIDGE CANYON
Durable Infrastructure

Natural bridges, such as the one south of Furnace Creek, are intriguing features that can form by many processes. The hike to Natural Bridge follows an easy and family-friendly gravel trail 0.3 mile up a dry streambed. It's best in the early morning or late afternoon when the high canyon walls provide shade.

Only 200 feet, or about 100 steps, from the information kiosk in the parking lot, you'll pass a low fault scarp to the right of the trail. It is one of many relatively youthful scarps along the western front of the Black Mountains, a part of the set discussed in Vignette 7.

The steep canyon walls along the first part of the trail consist mostly of reddish-brown conglomerate of the Mormon Point Formation, which is made up of gravel, silty mud, and ash derived from the Black Mountains volcanic field, all cemented together in a rock unit called fanglomerate—essentially a conglomerate made of alluvial fan deposits. Notice the high-angle faults that cut the fanglomerate in the canyon walls. The loose pebbles and cobbles beneath your feet are the tools that flash floods use to carve canyons like this one.

The floors of some of the side canyons are as much as 40 to 50 feet higher than the main canyon floor, so that their floodwaters cascade into the main canyon via vertical, smooth-sided chutes carved in the canyon walls. One of the best examples of a chute is about 100 yards before you come to the bridge. Several others are farther up the canyon beyond the bridge.

You won't have to stroll very far beyond the parking area and up the canyon before the impressive bridge comes into view, connecting both

GETTING THERE

The turnoff to Natural Bridge Canyon for southbound travelers is from the Badwater Road, 12.7 miles south of its intersection with CA 190 and 4.4 miles south of the entrance to the Artists Drive loop; for northbound travelers it is 3.5 miles north of Badwater. The steep 1.5-mile-long gravel road up to the Natural Bridge parking area is dusty and a little rough, with some dips and washboarding, but passable for standard automobiles. The hike to the bridge takes 10 to 15 minutes. See map on page 10.

A fault scarp cuts across the middle of the photo in alluvium at the mouth of Natural Bridge Canyon. —Photo by Jeff Knott

Hanging chute on the south side of Natural Bridge Canyon. Water and debris flows pouring out of side canyons during cloudburst storms erode these vertical chutes into the relatively soft fanglomerate. Because the mountains are rising rapidly relative to the valley floor, the main channel of the canyon is continually cutting down to match this uplift. Smaller side gullies have less erosive power and are left hanging.

sides of the narrow canyon. Once at the bridge, pause and marvel at the natural engineering that keeps the bridge in place, and try to deduce how it formed.

The opening beneath the bridge is 32 feet high. The bridge thickness is about 30 feet, its width is about 45 feet, and it is so cracked that you might wonder how much longer we must wait until it collapses. Although it appears fragile, the bridge looks almost the same today as in a photo from 1940, down to individual boulders and rugged rock projections.

Natural bridges, which cross waterways, and natural arches, which do not, form by various processes. In Arches National Park, Utah, arches develop in tall, thin sandstone walls that resemble shark fins. Intensely jointed (fractured) rock is overlain by tougher, erosion-resistant sandstone layers with a low concentration of joints. Erosion quarries out the jointed rock, leaving an intact roof. In Natural Bridges National Monument, also in geologically blessed Utah, bridges formed along a meandering, but deeply entrenched, tributary to the Colorado River when the river eroded through thin rock walls separating loops of the meanders.

Looking downstream at Natural Bridge in 2021 (left) and 1940 (right). Note the person in the 2021 photo at the same place the car was in 1940. There has been remarkably little erosion in the eight decades between the photos; individual rocks in the arch that were visible in 1940 are there today.
—George Willis Stose photograph from US Geological Survey photo archives

Neither of these processes explains Natural Bridge in Death Valley; groundwater was the culprit here. Natural Bridge, like several other such bridges in Death Valley National Park, consists of conglomerate, a tough rock. Tough layers in rock produce waterfalls, including the normally dry waterfalls that put an end to easy hiking in many desert canyons such as this one. Natural Bridge probably started out as a dry waterfall, but over time subsurface water in the gravel upstream found a way to seep through cracks in the tough layer, perhaps by dissolving soluble minerals or by plucking away one sand grain at a time in the porous sediment. This process led to the formation of underground channels that allowed increased flow of subsurface water. Muddy water initially carried silt and sand that eroded and widened the underground channels until eventually entire floods with sharp pebbles and cobbles like those you hiked on reamed out the channels even more to form the bridge. This process, known as piping, can have catastrophic consequences when it occurs beneath dams. Piping contributed to the calamitous failure of Teton Dam in eastern Idaho in 1976. Here in Death Valley, it led to a beautiful natural phenomenon.

Beyond Natural Bridge the first dry waterfall marks a profound geologic contact between greenish Proterozoic gneiss, about 1.7 billion years old, and fanglomerate, which is no older than a few million years. This contact is a major fault, a detachment fault, that helps accommodate extension across Death Valley. Here the fault is the surface running diagonally down from right to left. Green gneiss, below the fault, is visible through pervasive iron staining just to the right of the two people.

About 100 yards beyond the bridge, the canyon makes a broad turn to the right (east), and the canyon narrows. At 0.3 mile beyond the bridge, the rocks in the canyon walls change abruptly from Pleistocene conglomerate and other sedimentary rocks of the Mormon Point Formation to dark greenish-black Proterozoic gneiss. This distinction is not obvious, owing to pervasive red iron staining and mud washed onto the rocks, but it is profound. The contact between the two is a detachment fault, one of the best and most accessible exposures of one in Death Valley. Notice that 2 to 3 feet of the rocks above the fault are iron-stained red brown; the 1-foot-thick yellowish zone between them and the underlying gneiss is intensely crushed rock, indicative of the extreme deformation that has occurred along the fault surface. For more about this fault system, see Vignette 7.

The slippery gneiss is the first of several dry falls that make further passage up the canyon difficult, so most walkers turn back here. On the return walk, you'll have splendid views over Death Valley and Badwater Basin from the mouth of the canyon.

VIGNETTE 4

VENTIFACT RIDGE
Sand Blasts Stones in Mars-like Landscape

The wind often blows fiercely up and down Death Valley. In a 2003 report on Death Valley's climate, Steven Roof and Charlie Callagan noted that "wind may be the most underappreciated element of Death Valley's weather, at least until a visitor experiences a windstorm firsthand." Two unsheltered and barren ridges, lifted up along faults, lie directly in the wind's path between Furnace Creek and Badwater. We'll focus on the impressively severe erosion of surface stones by blasted windblown sand along the southern of the two ridges.

Ventifact Ridge consists of a coarse, bouldery accumulation of stones and sand swept down from mountains to the east. Most of the large stones littering the ground are chunks of black basalt, and many are riven by gas-bubble holes called vesicles. Wind has dramatically carved and polished the large basalt boulders on the crest of the ridge, changing their shape and size. Stones sculpted by sandblasting are called ventifacts, which means "wind-made," although "wind-shaped" would be more apt.

Ventifact Ridge is similar to a mound near the Artists Drive exit that is known informally as Mars Hill by NASA owing to its striking resemblance to parts of the surface of Mars. In fact, NASA tested its Mars Rover there in 2012. Ventifact Ridge may be even more Mars-like. Inspect a few stones on the ground. You'll notice that the surfaces of most of them exhibit signs of wear, and many surfaces have been abraded flat like facets on a gemstone. If two facets intersect, they can form an impressively sharp edge.

GETTING THERE

Ventifact Ridge is a low, narrow ridge that projects 1 mile into Death Valley southwest of the entrance to Artists Drive off Badwater Road, 8.5 miles south of California 190 and 7.8 miles north of Badwater. It rises a maximum of 150 feet above the surroundings. You'll find parking west of Badwater Road just south of Artists Drive entrance (don't confuse the site with Artists Drive exit, 3.7 miles north). It is an easy walk from the parking area to the top of the ridge, a climb of about 80 feet. Select a spur of the ridge with abundant boulders, and once on top, walk at least 300 yards southwest along the crest (36.3267, -116.8348). Remember that collecting rocks without a permit is prohibited in any national park. See map on page 10.

A storm-generated dust cloud bears down on Furnace Creek on April 27, 2016. Wind speeds are not archived at Furnace Creek, but half an inch of rain fell there that day. This windstorm blew out of the north, whereas Ventifact Ridge typically shows evidence for strong winds from the south.

View looking south down Badwater Road. Ventifact Ridge projects into the valley like a long finger and is subject to sandblasting from both north and south winds. The steep northern side is probably a fault scarp.

28   VIGNETTE 4: VENTIFACT RIDGE

Although the blue sky and humanoid may give away which of these is a photo of Mars and which is a photo of Death Valley, the environment on Ventifact Ridge is a good match for parts of the Martian surface. Both consist of angular blocks of basalt sitting on a rocky, sandy surface, and every large stone has been sandblasted.

Steep faces on larger stones commonly display pits that are more irregular, larger, and deeper than the original vesicles in the rock. Find a freshly fractured vesicular chunk of lava and compare the size, shape, and appearance of its vesicles with those on the exposed surfaces of nearby wind-abraded stones. On the wind-blasted rocks, the vesicles are larger, and the edges rounded. Many are elongated and some are joined together. Some large boulders with a near-vertical face have a pattern of grooves radiating in a half circle upward and outward. They look as if they've been eroded by a powerful fire hose, or maybe they're just having a bad hair day.

Radial grooves are produced by sand abrasion when powerful winds hit a flat, steep surface and radiate outward to pass around the obstruction.

Black basalt (left) and gray rhyolite (center) boulders with strong faceting from sand-blasting. These rocks have probably had half of their mass turned into sand and carried off by the wind. The Artists Drive area of colorful Cenozoic volcanic rocks in the background is the general source area for boulders on Ventifact Ridge.

All of these features—luster, facets, knife-edges, pits, grooves, and patterns—have been created by the blasting of wind-driven sand and silt. Sandblasting is a very effective method for cleaning soot-stained buildings; it is an equally effective erosion process. Look on the ground among the stones, and you will find little accumulations of well-sorted windblown sand forming tails to the lee (downwind) side of larger stones and in other sheltered spots. A large accumulation of sand downwind of a single large stone bears a spooky resemblance to a grave.

The ground-level outline of some of the large, stable boulders on the ridge suggests that sandblasting has worn away more than half the stone. How long would that take? No one knows for sure. Certainly centuries and perhaps millennia, but if you live in an arid region, don't underrate the power of wind as an erosive agent. For instance, if you were driving on a desert highway when a windstorm blasted a curtain of hopping sand grains in your path, it would take only a few minutes to render your windshield opaque. Windstorms that hit busy highways with drifting sand are bad news for insurance companies.

View looking southwest over a deep pile of sand trapped behind several 1-foot-long ventifacts near the west end of Ventifact Ridge. Most of the deep grooves are on the south sides of boulders, and most of the sand piles are on the north, indicating that the winds that do most of the damage come from the south. In the distance is Telescope Peak in the Panamint Range.

The view from the remote western end of Ventifact Ridge, an easy walk, is spectacular. To the south lies the brilliant white salt flat (Vignette 2), which extends eastward to the base of the Black Mountains at Badwater. Far to the south, you might glimpse the projecting profile of Mormon Point. Due east are the variegated, colorful, eroded slopes of the Artist Drive Formation. To the west are the huge alluvial fans at the base of the Panamint Range. Ventifact Ridge points almost directly at Telescope Peak, which graces the crest of the Panamints and is commonly snow-capped in the winter. Close by on the north is the back side of a smaller ridge parallel to Ventifact Ridge.

Strong winds frequently blow in the desert, and the paucity of vegetation works in their favor, allowing gusts to pick up and transport sand and silt. Consequently, ventifacts are abundant and widely distributed in arid areas. You may have noticed metal sheaths around the lower parts of wooden utility poles in desert areas. The metal protects them from being cut through by sandblasting. In some instances, erosion by windblown grit has left poles dangling on telephone or power lines after only a decade or two. If you happen to be on Ventifact Ridge wearing shorts in a strong wind, you may feel hopping sand grains attacking your legs. Not to worry, you won't be here long enough to become a ventifact, in spite of the efficiency of the sandblasting process.

VIGNETTE 5

ARTISTS DRIVE
A Garden of Earthy Delights

Although deserts are typically drab and sere unless wildflowers are in bloom, parts of Death Valley are flamboyantly colored. The Artists Drive area along Badwater Road looks like it was bombed with paint by aerial tankers. These colors radiate in the singularly named Artist Drive Formation, a mile-thick sequence of volcanic tuffs and lava flows with interbedded layers of sandstone and siltstone. These rocks underlie the slightly younger Furnace Creek Formation, an even thicker sequence of sandstone, siltstone, and mudstone layers with interbedded basalt flows and tuff. Isotopic dates indicate that the Artist Drive Formation was deposited between about 14 and 10 million years ago (middle Miocene). The Furnace Creek Formation is between 6.5 and 3.2 million years old (late Miocene and Pliocene).

From its junction with the Badwater Road, Artists Drive proceeds up the fan toward low hills backed by the steep western escarpment of the Black Mountains. Both the hills and the mountains are richly colored in browns and reds, and the rocks are distinctly layered. Although the layering is reminiscent of late Proterozoic to Paleozoic sedimentary rocks that are widespread in the park, the colors of these layered rocks mark them as volcanic. They are largely dark-brown, red, and gray basalt and andesite flows interbedded with light-colored tuff and fine-grained sedimentary rocks. You may see a dark basalt dike or two cutting the layers.

GETTING THERE

The entrance to the one-way road through the colorful Artists Drive area is along Badwater Road 8.6 miles south of its intersection with CA 190 at Furnace Creek. The Artists Drive loop road is paved but is also narrow, twisty, and not suitable for trailers. It goes counterclockwise, depositing you back on Badwater Road 4 miles north of the entrance, a half mile south of Mushroom Rock. Artists Drive is best traveled in the afternoon, when low sun over the Panamint Range bathes the slopes in a warm glow. See map on page 10.

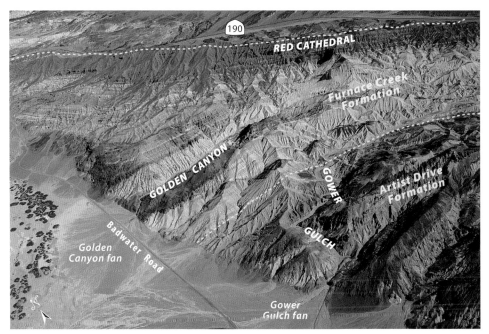

A thick succession of Cenozoic sedimentary rocks of the Furnace Creek and Artist Drive Formations crops out in the north, wedge-shaped end of the Black Mountains bounded by CA 190 and Badwater Road. In this aerial view looking northeast, the layers dip away and to the left (north) from the camera. The Furnace Creek Formation is mostly light-yellow sandstone and mudstone with interbedded dark fanglomerate, whereas the Artist Drive Formation consists largely of brightly colored tuff beds, tuffaceous sedimentary rocks, and dark lava flows. The contact between the two is marked by a dashed line. The sharply defined, linear range front between Golden Canyon and Gower Gulch is a young fault scarp. Zabriskie Point is just out of the photo to the upper right.

Age	Formation	Graphic Column	Rock Character	Thickness (feet)
HOLOCENE AND LATE PLEISTOCENE	unconformity		alluvium, alluvial fans, Lake Manly deposits	
MIDDLE AND EARLY PLEISTOCENE	Mormon Point		volcanic ash, conglomerate, mudstone	
PLIOCENE	Funeral		river gravel, interbedded basalt lava flows	1,000
MIOCENE	Furnace Creek		playa lake deposits, borates	5,000
	Artist Drive		river gravel, playa sand, siltstone, volcanic detritus	5,000
	TURTLEBACK	FAULT	granite intrusions, volcanic rocks	
PROTEROZOIC			gneiss	

A simplified stratigraphic column for the northern Black Mountains.

Bright blue-green tuff and tuff-rich sedimentary rocks of the Artist Drive Formation are widespread, as seen in this aerial view westward toward the valley floor of a remote area of the northern Black Mountains. Badwater Road crosses the upper center of the photo, and Artists Drive snakes through dark hills covered in basalt cobbles in the upper left.

Stop in the wide spot immediately after the road exits the hills and swings left (north), along the faulted mountain front. Get out of your vehicle and see how light-colored debris flows, which were shed westward from the mountains, have built a small bajada. This part of the road along the mountain front surfs the up-and-down topography of the bajada. Roadcuts in the banks of two deep gulches smooth out the bumps a bit and give you a good look at the innards of these fans. Pull off the road into the bottom of the second wash for a brief stop. You'll see cavernous weathering on the mountain front, but notice especially the huge boulders to the left, on and within the fan. Their size clearly indicates that it took more than a gurgling trout stream to transport and deposit them.

The gulches in this stretch of Artists Drive have incised deeply into the heads of their own fans. Fan-head incision occurs where the mountain, in which the feeder stream is located, is uplifted relative to the fan surface, or when the stream gradient and discharge abruptly increase. Look about 150 feet up the gulch and see the 6-foot-high dry fall in pink bedrock. This bedrock exposure is the fault scarp that lies along the break in slope between the mountain front and the alluvial fans.

VIGNETTE 5: ARTISTS DRIVE   35

The mid-section of Artists Drive loops around and into the heads of incised alluvial fans issuing from the steep front of the Black Mountains. A youthful fault scarp lies along the break in slope between the mountain front and the alluvial fans and is exposed here as a bright cliff running from just below center to above the road on the right.

The turnoff to the Artists Palette, a mile beyond the second set of roadcuts, offers a short side trip to an overlook of a colorful patchwork of volcanic rocks. Although the main geologic interest is in the colorful rocks, the wash in front of you is where R2D2 was captured by Jawas in *Star Wars: Episode IV*. After fully absorbing that factoid, look at the slope in front of you, tinted in shades of brown, pink, orange, green, yellow, red, aqua, ochre, hickory, fallow, citrine, sarcoline, eburnean . . . well, there are a bunch of colors, and the comparison to a paint-splashed palette in the hands of a beret-clad, smock-wearing artist is apt.

After leaving the Artists Palette, the road proceeds through a cut in fanglomerate deposits. It makes a deep U-turn and heads back toward the mountain front, twisting through dark lava flows and light-colored tuff, with local pockets of green (or seafoam) tuffs similar to those at Artists Palette. The road then winds through more fanglomerate deposits. Just as you exit the sinuous part of the road and head into its straight stretch leading back to Badwater Road, look at the cliff with a wave-cut base on the left overlain by some shingled gravel, both vestiges of the ancient Lake Manly shoreline.

Artists Palette in late-afternoon light. How many colors can you name?

GREEN ROCKS

Tuffs of various bright shades of green and blue-green are scattered throughout the Mojave Desert and eastern California, including Titus Canyon (Vignette 12) in Death Valley National Park, and Rainbow Basin in the Mojave Desert (*Geology Underfoot in Southern California*). The key to this beautiful coloration, and to the source of much of the color in any rock, is iron. Iron is a transition metal, a group of elements defined by particular electron configurations that allow them to take on more than one oxidation state. Iron gladly gives up either two or three electrons to oxygen, which is happiest when it has two extra electrons on board. A side effect of the electron configurations of transition metals is that they absorb parts of the visible-light spectrum, much more so than common rock-forming elements such as oxygen, silicon, and calcium, and the oxidation state has a large effect on which wavelengths of light are absorbed.

These ions of iron, Fe^{2+} and Fe^{3+}, impart different colors to minerals. Fe^{3+} generally adds red to brown colors—rusty colors, consistent with rust being a combination of various oxides and hydroxides of Fe^{3+}. In contrast, Fe^{2+} imparts a dark green to black hue, as in the common minerals olivine and pyroxene.

The green shades in tuff come from clays, which form from volcanic glass and feldspars by weathering and hydrothermal alteration. Most clays are whitish, but incorporation of small amounts of iron gives them color. Green clays seem to require a combination of Fe^{2+} and Fe^{3+}. The transition metal group includes metals such as copper, nickel, and chromium, but these are far less abundant than iron. Green colors are commonly said to indicate the presence of copper, but this is rarely true—usually it is just iron.

VIGNETTE 6

ZABRISKIE POINT, GOWER GULCH, AND GOLDEN CANYON
Diversionary Tactics

The law of unintended consequences took over in the 1940s when a little dynamite and a bulldozer were used to address the problem of infrequent but massive floods severely damaging the highway and hotel at the mouth of Furnace Creek Wash. If you drive into Death Valley by heading northwest on CA 190, notice how Furnace Creek Wash narrows dramatically and turns hard left as it passes the narrow notch adjacent to the hotel, and then emerges at the head of a barren alluvial fan. Although only a trickle of water usually flows in the wash, thunderstorms can unleash great floods that damage buildings and roads at the narrow bend. In 1941–1942, however, an apparent solution to this problem was engineered by diverting big Furnace Creek into little Gower Gulch, whose upper end nearly intersected Furnace Creek Wash.

Start your exploration of the diversion with a walk up the paved path to Zabriskie Point. It is de rigueur for all Death Valley visitors and geologists alike because the panoramic view is colorful and iconic, with sunrise an especially charming time. Your view is into upper Gower Gulch and the picturesque badlands carved into soft, fine-grained lake deposits of the 6-million-year-old upper part of the Furnace Creek Formation. The Panamint Range, from Telescope Peak on the left to broad Tucki Mountain on the right, form the backdrop.

The rocks around Zabriskie Point are relatively young and relatively undeformed—by Death Valley standards. In this area the rocks dip north (to the right), and so rock layers on the left are the oldest and those on the right the youngest. The craggy dark hills that blot out the Panamint Range on the left are volcanic rocks of the Artist Drive Formation of Miocene age. These lava flows and tuffs were submerged by a lake that deposited the light-yellow siltstone, now eroded into the badlands around Gower Gulch, including the prominent spire of Manly Beacon. The dark-brown layer about 1,000 feet south of the viewpoint is a basalt flow that invaded the lake—a last gasp of Artist Drive volcanism.

Reddish cliffs known as the Red Cathedral, on the right side of the panorama, consist of conglomerate layers from alluvial fans that flowed

VIGNETTE 6: ZABRISKIE POINT, GOWER GULCH, AND GOLDEN CANYON

GETTING THERE

This vignette explores three parts of the Furnace Creek Wash–Gower Gulch drainage systems: the diversion of Furnace Creek into Gower Gulch at Zabriskie Point; the channel of Gower Gulch from Zabriskie Point to the west foot of the Black Mountains; and the Gower Gulch alluvial fan on the Death Valley floor. First, inspect the diversion at the headwater of Gower Gulch by driving 3.6 miles southeast up Furnace Creek Wash on CA 190 from its intersection with the Badwater Road to the wide parking area at Zabriskie Point.

A scenic and geologically diverse 5-mile loop trail goes up Golden Canyon, crosses into the north fork of Gower Gulch, and descends the gulch to the mountain front where a trail leads back to Golden Canyon. Enthusiastic hikers may enjoy a hike from Zabriskie Point down upper Golden Canyon and a connection to the main channel of Gower Gulch. This colorful trek requires a bit of easy scrambling but reveals a number of interesting sedimentary features including polished fault surfaces and beds with mud cracks, ripple marks, and raindrop prints.

You can reach the lower end of the Gower Gulch system by taking the Badwater Road 2.6 miles south from its intersection with CA 190 to a section of the road that spans a broad gulch 0.6 mile south of Golden Canyon and 1.8 miles north of Mushroom Rock. Park in a wide spot west of the highway and north of the dip, and walk up the broad stream channel cut into the alluvial fan.

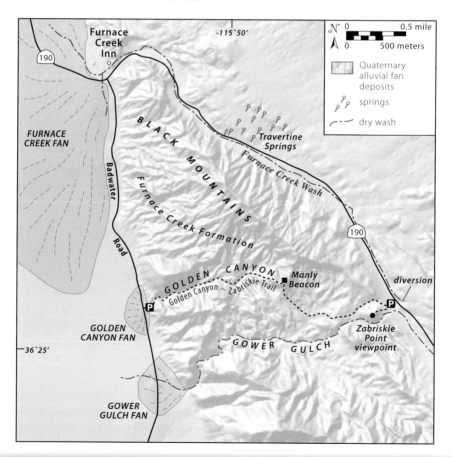

into the lake basin, much as alluvial fans in Death Valley are trying, but failing, to fill in that huge basin. Look closely and see that the yellow siltstone and darker conglomerate beds are interlayered. It may not look much like it now, but this was an old shoreline where lake sediments interfingered with alluvial fan sediments, equivalent in geologic evolution to the current shoreline around the alluvial fan at Badwater (Vignette 1). The toe of the alluvial fan built out into the lake, and then when the lake rose, perhaps because of a period of increased rainfall or a lava flow damming its outlet, lake sediments were deposited upon the conglomerate. Then the lake level dropped again so that the fan built out into the basin again. Repeat as necessary.

Gower Gulch is but one of several short, narrow canyons eroded into the lakebeds. It lies within the narrow wedge of the northern Black Mountains that is bordered on the northeast by Furnace Creek Wash and on the west by the floor of Death Valley. These short canyons drain westward from a headwater divide, which, at Gower Gulch, used to lie unusually low and close to Furnace Creek Wash. Each gully has built a small alluvial fan on the Death Valley floor, and Badwater Road traverses the toes of these fans, which are cut by obvious fault scarps.

Sometime in 1941 or 1942 a mining company bulldozed a little diversion dam across Furnace Creek Wash and dynamited a bit of yellow rock at the head of Gower Gulch, so that floodwater from upper Furnace Creek Wash would make a sharp left turn at the dam and flow down into the much smaller Gower Gulch. The diversion had drastic and unanticipated, yet predictable, effects on both drainages. For geologists, it was a fascinating experiment executed at full scale in a natural setting, but it produced several headaches for the National Park Service including serious erosion of CA 190 near the diversion, damage to natural springs and native mesquite in lower Furnace Creek Wash, changes in the level of groundwater beneath Furnace Creek Village, dissection of the Gower Gulch fan, and repeated damage to Badwater Road.

Have a look at the diversion by walking down from Zabriskie Point to the southeast edge of the parking area (left side as you're driving in), where the channel of Furnace Creek Wash diverges from its normal course and plunges through a steep and narrow gorge into the head of Gower Gulch. With this diversion, Furnace Creek Wash dumps its discharge of floodwater and debris into the much smaller Gower Gulch system.

Although the diversion is artificial, accomplished with bulldozer and dynamite, the original low, soft bedrock divide between Furnace Creek and the head of Gower Gulch assured that a natural diversion would be inevitable within a geologically short time, perhaps a century or so. It could have been accomplished either by headward erosion of Gower Gulch, or by accumulation of gravel in Furnace Creek Wash up

VIGNETTE 6: ZABRISKIE POINT, GOWER GULCH, AND GOLDEN CANYON

The panoramic view of Death Valley from Zabriskie Point is iconic. Rocks dip to the right (north), so the dark volcanic rocks at left (Artist Drive Formation) are older than the yellow siltstone beds (Furnace Creek Formation), which interfinger with the brown conglomerate beds on the right.

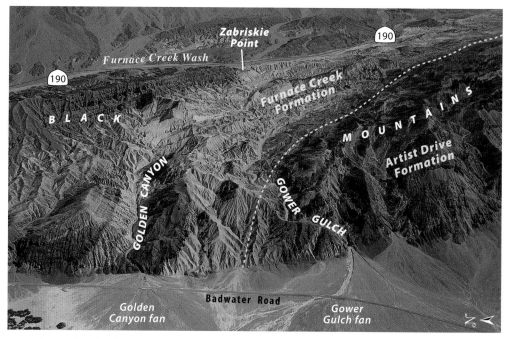

Aerial view looking east up Golden Canyon and Gower Gulch toward the Zabriskie Point viewpoint and the artificial connection between Gower Gulch and Furnace Creek Wash. The Badwater Road stretches along the lower part of the photograph. Prior to diversion, Furnace Creek drained an area eighty-five times larger than that of Gower Gulch. Now all but the largest floodwaters of Furnace Creek funnel down into the smaller Gower Gulch.

VIGNETTE 6: ZABRISKIE POINT, GOWER GULCH, AND GOLDEN CANYON   41

The Furnace Creek diversion into Gower Gulch in 2021, looking southeast from near the Zabriskie Point parking area. The parking area and highway are at the pre-1941 level of Furnace Creek Wash, and all the erosion below that level has occurred since then. The current main channel of Furnace Creek comes toward the camera parallel to the highway, turns abruptly, and drops into Gower Gulch gorge at lower right. Headward erosion of the tributary gully on the left has repeatedly undermined the highway. The diversion of Furnace Creek solved the erosion problem at the mouth of the wash but created a bad one here.

to a level where the wash would overtop the divide into Gower Gulch. Natural diversions occur commonly as adjacent drainage systems fight to increase their territory. A diversion nearly always results in unusual channel configurations and anomalous landforms until the system smooths them out.

From the Zabriskie Point area, Furnace Creek drops about 670 feet over 3.5 miles (3.6 percent) to the range front at the Furnace Creek Inn. The diverted stream course drops 800 feet in 2.5 miles to the range front at Gower Gulch, a gradient nearly twice as steep (6.1 percent). Once Furnace Creek Wash floods started flowing down the considerably steeper route via Gower Gulch to the Death Valley floor, they easily eroded the soft Furnace Creek Formation and were soon entrenched in the new course, carrying coarse debris to the tip of its alluvial fan in Death Valley.

The Furnace Creek Wash drainage is much larger and more capable of carrying floods than the Gower system. Before diversion, Gower Gulch probably carried runoffs of mostly a few tens of cubic feet per second. After diversion, floods that discharge from a much-enlarged drainage area may range up to hundreds to even thousands of cubic feet per second. One 1968 storm generated a flood estimated at 7,000 to 10,000 cubic feet per second. That is equivalent to low-level natural flow of the Colorado River through the Grand Canyon, and the mass of water was roughly equivalent to the mass of about five to seven Olympic-size swimming pools flowing by each second. That is a lot of destructive power.

Floodwaters carrying cobbles and boulders cut through the soft siltstone like a chainsaw, rapidly deepening the narrow gorge. How far the chainsaw has cut down since 1941 is difficult to estimate because the

Looking upstream from above the slot carved by diversion of Furnace Creek Wash. The slot is 41 feet deep here, much of which has been carved since 1941.

original configuration of the drainage divide can only be estimated; we put the number at 30 feet in the 80 years since the diversion happened.

Now that you've seen the hydraulic disarrangement at the top of Gower Gulch, go to the lower end of Gower Gulch to see its alluvial fan and incised channel. The walk from the Zabriskie Point parking lot down to the valley floor is quite fun, involving a drop of 800 feet over a distance of about 2.5 miles. Alternatively, follow directions in Getting There to where Gower Gulch crosses the Badwater Road 2.6 miles south of its junction with CA 190.

Walk a few hundred yards eastward up the channel from Badwater Road. Gower Gulch has cut a channel 5 to 6 feet deep here. Notice how the bank height increases from zero at the road to 12 to 15 feet at the mountain front a quarter mile upstream. Notice also how the presently active channel broadens and shallows downslope so that it eventually intersects the surface of the original Gower Gulch fan. There it spreads into a network of distributary channels from which it deposits debris on the old fan, simultaneously building the toe of the fan outward, burying more and more of the old fan, and eventually burying it completely—and the Badwater Road as well. Here the fan is building out over the lake deposits at its toe, akin to what we saw in the red cliffs at Zabriskie Point.

The active channel has scoured 20 feet into the old fan surface by high-volume storm flow since the diversion of Furnace Creek in 1941–42. The walls of the wash are eroding rapidly; large chunks of poorly cemented fanglomerate collapse into the channel, but these typically wash away within several years. Snow caps Telescope Peak on the crest of the Panamint Range.

The channel floor is dominated by coarse Furnace Creek gravel—worn and somewhat rounded cobbles, mainly carbonate rocks, with a few clusters of boulders greater than 10 inches diameter. You won't see anything comparable in the walls of the new channel as you walk from the toe of the fan to the mountain front—neither the sizes, the rounding, nor the types of the cobbles match. The prediversion alluvial fan was built entirely of rock eroded from the Black Mountains. The modern stream channel, which is currently cutting down into the prediversion fan at an average rate of a few inches per year, carries much coarser cobbles, as well as an occasional boulder of limestone—a rock that is found several miles away in the Funeral Mountains to the northeast of Furnace Creek Wash and CA 190. These large stones are indicative of much stronger stream flows that came down from Furnace Creek.

The first major postdiversion flood onto the fan from the mouth of Gower Gulch probably resembled the stream from a fire hose. It quickly carved a deep channel in the fine-grained alluvium at the head of the fan so that subsequent floods were confined by the incised channel until they were well down the fan.

The diversion causes headaches for National Park Service crews maintaining Badwater Road, and road damage will continue to worsen into the foreseeable future. Every significant storm leaves an accumulation of debris on the highway, and accompanying stream scouring requires repair work to restore the roadbed. In a final irony, large floods mostly bypass Gower Gulch. Seventy percent of the floodwater in Furnace Creek Wash in 1968, 1977, 1985, 2004, and 2015 bypassed the diversion and took the original course out onto the Furnace Creek alluvial fan, causing extensive damage, including two fatalities in 2004.

VIGNETTE 7

DANTES VIEW
It's Turtlebacks All the Way Down

One of the most spectacular views in the California desert and a sublime place to watch the sun rise or set is Dantes View on the crest of the Black Mountains. The viewpoint is a vertical mile above Badwater and looks across the full width of the salt pan to the Panamint Range. The airy view gives the feel of being in an airplane even though your feet are firmly planted on the ground.

Dantes View has long been a popular tourist destination, but it also is a must-do geologic stop because it poses a great mystery: a stack of sedimentary rocks fully 5 miles thick has gone missing. They are present in the surrounding ranges but almost entirely missing here. Where did they go? The answer, or at one of several possible answers, is best appreciated from the viewpoint.

Before driving the vertical mile up to Dantes View (don't worry, it's not literally vertical!), let's review the relevant rock units that are part of the story. The oldest are Proterozoic metamorphic and igneous rocks, 1,700 to 1,400 million years old. They make up the rock upon which late Proterozoic and Paleozoic marine sedimentary rocks were deposited at intervals between about 1,000 and 250 million years ago. These sedimentary rocks accumulated to remarkable thicknesses. How did a stack of sediments 5 miles thick get deposited in the ocean?

A billion years ago most of the continents were amalgamated into a supercontinent known as Rodinia. California, Nevada, and surrounding states were in the middle of this supercontinent, but about 800

GETTING THERE

Dantes View is reached via a 13-mile spur road that takes off from CA 190, 12 miles southeast of Furnace Creek and 3 miles northwest of the park boundary. Although the road starts out with a gentle climb, it steepens, then narrows to one lane. Trailers are not allowed on the switchbacks beyond the narrows. The parking area lacks facilities, although pit toilets are at the trailer parking area below the switchbacks. The turtlebacks and recent fault scarps that are part of the Dantes View story can be studied along Badwater Road south of CA 190. Vignettes 1 through 5 (see map on page 10) discuss sites along Badwater Road.

million years ago Rodinia rifted apart, breaking into several continents. The pieces that would later become Australia, eastern Antarctica, and southern China rifted away from a large landmass that included most of present-day North America, and an ocean developed between them and steadily widened. California faced this ancient proto-Pacific Ocean. Rivers brought sediment to the sea, beaches formed, and sandstone, shale, and limestone were deposited offshore.

Dantes View (5,440 feet) looks across Badwater Basin (-282 feet) to Telescope Peak (11,048 feet). The apex of the huge Trail Canyon fan, below and to the right of Telescope Peak, is 2,400 feet above sea level; the apex of the fan just south of Badwater, in the foreground, is approximately at sea level.

The ancestral West Coast of this time was a dull place. Land life consisted of little more than bacterial films on rocks, and oceans contained only microbes and various jellyfish-like creatures of which little fossil record is left. A vacation to the coast of proto-California would have been boring, and probably smelly as well. Oxygen levels were low, so you would have needed an oxygen mask, and there was little ozone in the upper atmosphere to block ultraviolet rays, so you would have wanted to stand in the shade of a tree—except there were no trees. Imagine the East Coast of today, but with no land animals, land plants, fish, hamburger joints, condominiums, casinos, and the like.

The shoreline at that time ran southwesterly from Utah to the Mojave Desert. It can be traced by following ancient beach sand deposits (now sandstone) that are in Utah, Arizona, Nevada, and California. These rocks are famously exposed in the Grand Canyon. The oldest flat-lying sedimentary rock there is a beach sand, the Cambrian Tapeats Sandstone, and successively higher layers represent marine sedimentation, punctuated by periods of erosion, as sea level rose and fell. The rocks at the canyon rim are the Permian (latest Paleozoic) Kaibab Limestone, and the entire sequence is about 3,500 feet thick.

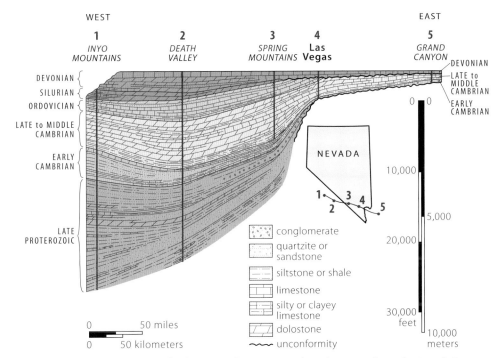

Summary of thickening of Paleozoic sedimentary rocks going west from the Grand Canyon to Death Valley and the White-Inyo Range. The thin part on the east was deposited on and near the old continental shelf, and the deeper section to the west farther offshore.
—Modified from Stewart and Poole, 1974

Deeper water lay to the west, and the ocean floor under it sank slowly, both by thermal contraction of the newly formed oceanic crust and under the weight of the accumulating sediments. This allowed thicker and thicker sequences of rocks to accumulate, and about a hundred or so miles offshore, in the area of what is now Death Valley, the rocks accumulated to more than 5 miles thick, even though the sea was never more than a few thousand feet deep. The mountains of Death Valley consist largely of this enormously thick stack of sedimentary rocks. The Bonanza King Formation alone, which was deposited during the latter part of Cambrian time, is as thick as the entire Paleozoic section in the Grand Canyon.

A Trip to Dantes View to See the Problem

From Furnace Creek, CA 190 climbs past Zabriskie Point through pale sedimentary rocks of the Furnace Creek Formation of Miocene age. The mountains on the skyline to the northeast are the south end of the Funeral Mountains with its many layers of gray Paleozoic sedimentary rocks that swoop up into the air, part of the 5-mile-thick stack. After turning south on the road to Dantes View, you'll see Artist Drive Formation volcanic rocks on both sides of the road, some of them stained various shades of yellow and orange by hydrothermal alteration.

Reddish volcanic rocks of the Artist Drive Formation and the Billie Mine at Ryan stand out in front of a swooping, well-layered sequence of Paleozoic sedimentary rocks of the southern Funeral Mountains on the skyline. A similar stack of these Paleozoic rocks, which should be on top of the Black Mountains, is missing.

Peaks in the Panamint Range, with a few of the higher Sierran peaks visible through low points in the ridge. Split Mountain is 110 miles away.

About 3.5 miles beyond the turnoff from CA 190, Dantes View Road turns west, enters volcanic rocks, and begins its steady climb to the parking area at the road's end. Finally, upon arrival at Dantes View parking area, which is built on Miocene volcanic rocks, you'll see Proterozoic gneiss and other metamorphic rocks visible a few ridges to the north. Where is the stack of Paleozoic rocks that occurs throughout the Death Valley region? Let's take in the view while we search for the missing rocks.

Dantes View, more than 1 mile (5,440 feet) above sea level, looks down the precipitous and craggy western slope of the Black Mountains to Badwater (282 feet below sea level). Across Death Valley looms the Panamint Range, capped by Telescope Peak, and if you know where to look to the right of Telescope Peak, you can spot several 14,000-foot peaks in the Sierra Nevada, including Mt. Williamson, Split Mountain, and North Palisade, 117 miles away. Mt. Whitney (14,505 feet), the tallest peak in the contiguous United States, is barely hidden about one finger (held at arm's length) to the left of Mt. Williamson. On a clear day, peaks in the San Gabriel and San Bernardino Mountains, 145 miles away, may also be seen. It is remarkable that the highest peak in the lower 48 states, in the landscape in front of you, and the lowest point in North America, below you, are in the same county—Inyo County. If you sawed off Mt. Mitchell, the highest US mountain peak east of the Mississippi, at its base, you'd have to stack it up twice to reach the top of Telescope Peak.

It may not be especially obvious when you gaze across the valley that the Black Mountains, which host Dantes View and culminate in Funeral Peak (6,384 feet), are only half as high as the Panamint Range. Unless you drive the West Side Road at the base of the Panamints, the scale of the alluvial fans spilling from the range is difficult to appreciate. Trail Canyon and Hanaupah fans across the valley stretch about 6 miles from their apexes to their toes on the valley floor, ten times as far as the Badwater fan on the valley floor below you, and their volumes are around one hundred times greater.

Here on Dantes View, you are standing on a great mystery, one that confounded geologists into the 1980s and still puzzles. The Panamint Range across the valley contains a huge, 5-mile-thick stack of Proterozoic and Paleozoic sedimentary rocks. So do the Funeral Mountains to the north, and the Resting Spring and Nopah Ranges to the east. You would think Paleozoic rocks ought to be here on the Black Mountains, too, but a similar stack is almost entirely missing. Where is it? Was it eroded away completely from the Black Mountains? If so, why aren't a few uneroded scraps still here? Why are they still present and so thick in nearby ranges? Were they never deposited where the Black Mountains now exist?

The magnificent view from Aguereberry Point in the Panamint Range is a must-see for any Death Valley aficionado and is complementary to the vista from Dantes View. The view is to the southeast across Trail Canyon and the salt pan to Mormon Point at the right (south) end of the Black Mountains. All the layered rocks in the foreground are Paleozoic sedimentary rocks, chiefly the Carrara and Bonanza King Formations—the thick sequence that is missing in the Black Mountains. The 6-mile dirt road to Aguereberry Point is generally passable to 2WD cars, but moderate to high clearance is necessary.

This paradox puzzled geologists for decades. Around 1980, though, a possible answer emerged with recognition that normal faults can operate at low angles as well as steep ones. Prior to this realization, low-angle normal faulting had been thought to be mechanically impossible, because the upper plate should fall apart along preexisting fractures when stretched. Field evidence that such low-angle faults actually exist was widely recognized in the latter part of the twentieth century.

And this recognition provided a solution to the mystery of the missing rocks. When you stand at Dantes View and look across the chasm at the Panamint Range, you are looking at the answer: the Black Mountains were, in a sense, pulled out from beneath Paleozoic rocks now in the Panamint Range. The rocks are not missing, they are right in front of you but are now over 10 miles away. This gross simplification of the true three-dimensional complications of such tectonic events distills the basic answer—or at least one answer, because there is still disagreement on this topic.

With this palm-to-forehead realization, many pieces of Death Valley geology fell into place. In 1966 Clark Burchfiel, of Rice University and later Massachusetts Institute of Technology, and Jack Stewart of the US

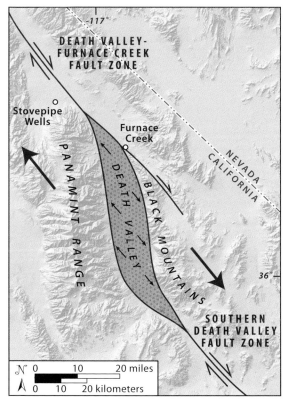

In the pull-apart origin of Death Valley, the right-lateral Death Valley fault zone has a north-south jog that departs from its overall northwest-southeast orientation. Slip on the fault system opens up a gap along the north-south segment that is Death Valley.

Geological Survey had proposed that Death Valley is a pull-apart basin formed by a large bend in a strike-slip fault system. A right-slip fault system comes southeastward from northern Death Valley to Furnace Creek, where it swings almost due south along the front of the Black Mountains to Shoreline Butte, and then resumes its southeastward course and exits the south end of the valley, thus making a slanted Z shape.

If you sketch a Z shape on a piece of paper, cut along the line, and then slide the pieces parallel to the upper cut with your hands, you'll open a chasm along the central cut. This is a simplified view of the origin of the Death Valley trough, although three-dimensional details are not addressed in this simple two-dimensional model. Later Stewart applied his knowledge of the thickness and stratigraphic variations of Proterozoic and Paleozoic sedimentary rocks to test the pull-apart hypothesis. He noted that because this succession of rocks thickens dramatically from the Grand Canyon to Death Valley, the amount of offset on the strike-slip faults can be estimated by matching up lines of equal thickness of a given formation. He came up with an estimate of about 50 miles of right-lateral slip.

Looking north from Dantes View into northern Death Valley. The craggy ridge in the middle ground is mostly Proterozoic gneiss in the Badwater turtleback; the reddish rocks in the foreground and behind the gneiss ridge are Cenozoic volcanic and sedimentary rocks that make up the range front all the way to Artists Drive and Furnace Creek. Distant ranges (Panamint Range, left; Funeral and Grapevine Mountains, right) have thick sequences of Paleozoic sedimentary rocks that are not present in the Black Mountains.

That is a surprisingly large amount, much greater than the width of the Death Valley trough. Stewart realized that sliding the strike-slip faults back would cause the Black Mountains and Panamint Range to overlap. He solved that problem by restoring the Panamint Range to a position above the Black Mountains. The mystery of the missing rocks was solved. Thus, when you are at Dantes View, you are looking at rocks in the Panamint Range that were transported to the northwest, off the Black Mountains, on a major detachment fault.

This solution also explained another mysterious feature of Death Valley—the miles-long dome-like structures of metamorphic rocks that plunge beneath the valley floor on the west front of the Black Mountains. These structures are known as "turtlebacks" owing to their fanciful resemblance in shape to a turtle's shell. The turtleback landforms are best admired along the Badwater Road.

Turtlebacks and Fault Scarps along the Black Mountains Front

The drive south along Badwater Road from Furnace Creek and CA 190 goes through a wonderland of faults and fault scarps, and before you visit the turtlebacks, you ought to examine some of these beautiful little rips in the landscape. Stop at a gravel parking turnout on the west side of Badwater Road, 0.7 mile south of CA 190 (36.4368, -116.8515). Northbound travelers reach the parking spot 1.3 miles north of the turnoff to Golden Canyon. A prominent low fault scarp is parallel to the road and here only 10 yards east of it. Many gullies scar the scarp face, and the larger ones have built small alluvial fans at the scarp's base.

This scarp is geologically quite young—to a geologist it looks like it formed yesterday—but the lack of rain keeps erosion to a minimum, thus its youthful appearance is misleading. Earthquakes large enough to cause ground breakage have not happened in Death Valley in the last few hundred years, so the scarps are older than that. As you proceed south along Badwater Road, you'll see that the scarps such as this one cut across a succession of small alluvial fans, including one at the Natural Bridge parking area, almost all the way to Badwater and beyond.

Little fault scarps such as these are common in Death Valley, and once you know what they look like you'll probably notice many more. More impressive faults with much greater displacement are in the valley, though, and you can stand right on some of them, too.

Drive south on Badwater Road to Badwater, passing the sites of Vignettes 3, 4, and 5, and note your mileage at Badwater. The road wraps around the fan at Badwater and loops around six more fans before pointing back at the mountain front 10 miles from Badwater. At this point, you are aimed southeast directly at the pyramid-shaped Copper Canyon turtleback. Find a convenient place to park and take in the view.

From this viewpoint you are looking up the long axis of the turtle's back. The right (southwest) and left flanks fall off steeply, but the nose plunging toward you is gentler. The road here is about 200 feet below sea level, and the highest part of the turtleback visible from here is 5,000 feet above sea level, so at 1 mile high, this is a big turtle. The Panamint Range behind you was pulled off this huge fault surface, exposing these mid-crustal metamorphic rocks. Turn around and look at the Panamints, whose crest is about 4 miles west of you, to get the full impact of all that activity. If the fault slip happened at about a tenth of an inch per year (see below), then 2 million years ago the rocks at the top of the turtleback would have been a mile or so under your feet.

The near side of the Copper Canyon turtleback has a thin veneer of pinkish rock about halfway up, a weak Proterozoic-age marble that was smeared out along the fault surface. Lower down, coppery-red sedimentary and volcanic rocks wrap around the nose, in fault contact with the Proterozoic rocks. They were deposited at the surface, whereas the rocks of the turtleback have a more complex history, having been metamorphosed at depths of perhaps 10 miles or deeper. The faulting has brought these two parts of the crust together. You may put your finger on a turtleback surface in Natural Bridge Canyon (Vignette 3).

Fault scarps cutting young alluvium in the first few miles south along Badwater Road from CA 190 look considerably younger than the first author, who is standing on them. Here, looking north, the scarp is about 100 feet west of the low mountain front.

Although the sunlit Copper Canyon turtleback looks like a symmetric dome in this photo, the turtleback surface stretches away from the camera to its summit fully 1 mile above the valley. Pink rocks on the left side of the dome are smeared-out marble, and reddish rocks in lower left are tilted Miocene volcanic and sedimentary rocks in the upper plate of the turtleback fault. Displacement on this fault is probably measured in tens of miles.

Closer view of the fault at the nose of the turtleback, with bright red rocks above the fault and greenish ones below.

The Mormon Point turtleback, south of Copper Canyon and north of Shoreline Butte, is another huge arching structure that plunges northwestward under the valley. A small gully that drains its western side exposes a detachment fault that puts fanglomerate and debris flows onto Proterozoic gneiss, as in Natural Bridge Canyon. To find the gully, proceed south around the broad arch of Mormon Point and watch for the brown Mormon Point sign. Drive south 0.6 miles and watch for a south-facing diamond-shaped warning sign on the left (east) side of the

The south wall of the canyon on the west flank of the Mormon Point turtleback (36.0459, -116.7606) is mud-stained to a uniform color, but rock below the white dashed line is brecciated Proterozoic gneiss and rock above it is faulted young fanglomerate and debris flows. Faults cutting the fanglomerate (red dashes) do not cut the master low-angle fault.

road. Drive 0.1 mile farther, park at 36.0451, -116.7620, and walk about 150 yards up the canyon to the east. There you will see the low-angle fault and dozens of high-angle, curving faults that reach down to the low-angle fault but do not cut it. As at Natural Bridge Canyon, this is likely the fate of the small faults cutting alluvium south of Furnace Creek.

Global Positioning System (GPS) instrumentation has revolutionized the study of tectonics, and continuous measurements are being recorded from instruments all over the area covered by this book. Two stations are in the Panamint Range and another is in the Black Mountains. A station near Ballarat in the Panamint Range is moving northwest relative to stable North America at about 0.28 inch per year, and a station east of Mormon Point in the Black Mountains is moving about 10 degrees more northerly at 0.16 inch per year. If we look at the difference between these two movements—that is, how the Panamint station is moving relative to the Black Mountains station—we get about 0.13 inch per year, toward an azimuth 15 degrees west of northwest. This azimuth is roughly parallel to the long axes of the three turtlebacks. They appear to be huge corrugations in the detachment fault along which the Panamint Range was transported northwestward, exposing the Black Mountains to daylight.

VIGNETTE 8

LAKE MANLY
A Huge Bathtub without a Drain

Let your imagination picture a lake hundreds of feet deep and 100 miles long occupying Death Valley thousands of years ago. Take an imaginary boat cruise from one end to the other. The north, south, and west shores would be irregular with inlets, peninsulas, and offshore shallow water. The east shore along the Black Mountains would be steep, relatively straight, and the water would be mostly deep. You could almost reach out and touch the Copper Canyon and Badwater turtlebacks from your imaginary boat. Furnace Creek Wash would form an interesting inlet. But if you see such a lake in Death Valley today, it is probably just a heat-shimmering mirage on the dry desert floor.

The floors of some of the Great Lakes, Lake Chelan in Washington State, and large lakes in Canada's Northwest Territories are farther below sea level than Death Valley, but they are filled with water. Death Valley once held such a lake, called Lake Manly after one of the early pioneers to enter the valley. It was about 600 feet deep and close to 100 miles long. That was during an ice age when the climate was colder and wetter and glaciers occupied mountains within the greater Death Valley drainage system. Lakes formed and disappeared repeatedly in Death Valley over a period possibly exceeding 1 million years. Lake Manly is the name applied to all Pleistocene water bodies older than about 10,000 years in the valley. Their various phases are called stands.

Just as water levels in your household bathtub can leave rings that remain until you scour them away, so also did ancient Lake Manly leave remnant rings, called strandlines, on Death Valley's tub. They consist of wave-cut cliffs and benches, deposits of tufa (calcium carbonate precipitated from lake water—see Vignette 17), and accumulations of shoreline gravels. Such gravels commonly contain smooth, well-worn, tabular pebbles, flattened by sliding up and down beaches in the swash of breakers. Like any good housekeeper, nature busily scours away the bathtub rings with erosion every time a stand of a lake dries up. Consequently, strandlines have been fully erased in many places and are only faintly visible in others.

A set of Lake Manly strandlines, which may provide you with the most convincing evidence that a big lake once occupied Death Valley,

58 VIGNETTE 8: LAKE MANLY

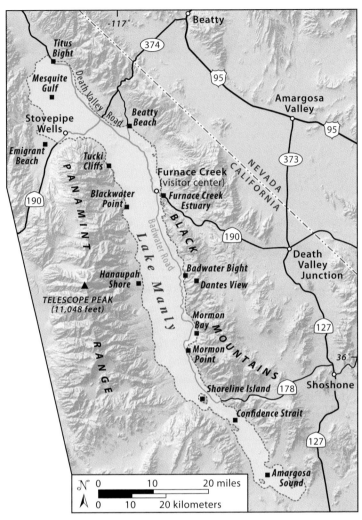

Map reconstruction of the extent of Pleistocene Lake Manly at maximum 600-foot depth. We added some fictitious names to help you visualize the lake and your imaginary end-to-end cruise.

GETTING THERE

Evidence for a former large lake in Death Valley is mostly faint, fragmental, and scattered, but such features are easy to see once you know what to look for. We recommend three places. In the southern part of Death Valley, 1 mile west of Ashford Mill, Shoreline Butte's many ancient shorelines on its basaltic slopes provide convincing evidence of the lake's former presence. The steep western front of the Black Mountains, at Badwater and for half a mile northward, displays horizontal shoreline features of cemented beach gravel, some more than 300 feet above the base of the mountains. North of Furnace Creek, the Beatty Cutoff Road to Daylight Pass bisects a large beach ridge 1.8 miles northeast of its junction with California 190.

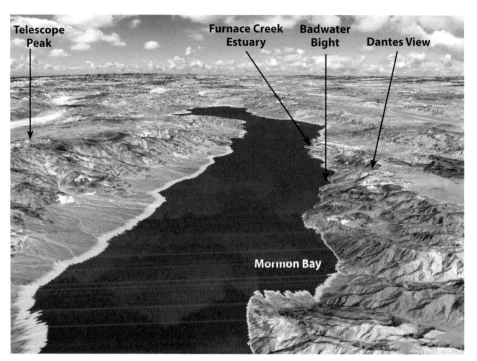

View of the Death Valley trough, looking north, as it might have looked when Lake Manly was at its maximum level (you will have to use your imagination to fill in appropriate vegetation, which was abundant in the cooler, wetter climate). Beaches on the west-side fans were broad, whereas the eastern shore near Badwater was bounded by precipitous cliffs. Furnace Creek emptied into an estuary above the present site of the Furnace Creek settlement.

is prominent on the flanks of Shoreline Butte in the valley's south end. Especially large waves, generated by storm winds with a long overwater fetch, vigorously attacked the butte's north and northeast flanks, creating well-defined strandlines there. Strandlines in the butte's relatively tough 1.5-million-year-old basaltic lavas have resisted nature's scouring rather well. They are more obvious when backlit by the afternoon sun. The times of these stands are difficult to determine, but the most prominent of them formed probably about 150,000 years ago, around the time of the Tahoe glaciation but considerably before the latest (Tioga) glacial advance in the Sierra Nevada. Younger, much smaller, short-lived lakes have likely occupied Death Valley during the last few thousand years.

Lakes that have stable overflow outlets and sufficient inflow to continually overflow at the outlet form few but strong strandlines, but such was not the situation at Lake Manly. Lake Manly had no outlet; it was a bathtub without a drain because there was no place lower for the water to go. During periods of wet climate, the water level rose; during

The horizontal streaks on the northeast flank of Shoreline Butte, as viewed from the Badwater Road, are Lake Manly strandlines, carved by waves battering what was then Shoreline Island.

dry times the water evaporated, and the level fell. The large number of strandlines indicates that water depth fluctuated frequently, seldom stabilizing long at any level.

Strandlines on Shoreline Butte tell us something about the maximum water depth in the late stands of Lake Manly. The butte's summit is 663 feet above today's sea level. The highest confidently identified strandline is at 285 feet above sea level, although tantalizingly faint suggestions of still higher strands exist. The highest strandline on Shoreline Butte is therefore about 570 feet above the lowest point on the valley floor, 282 feet below sea level. Over the lowest part of the valley near Badwater, strandline elevations on the western front of the Black Mountains also suggest a lake close to 570 feet deep. That depth seems reasonable, but faulting has complicated the estimates of maximum lake depth, so geologists hedge by saying that the maximum depth of Lake Manly was probably between 500 and 600 feet.

The steep mountain front in the deep embayment that creases the Black Mountains north of Mormon Point retains faint horizontal marks of Lake Manly strandlines, 200 to 300 feet above the valley floor. Most of the marks are carbonate-cemented gravels that form horizontal streaks along the mountain front, like those near Badwater. Remnants of young (post-lake) fault scarps 75 feet high along the base of the Black Mountains indicate that the lake may have been at least 75 feet shallower than the Black Mountains strandlines suggest.

Clear evidence that a lake once filled Death Valley is given by well-developed beach ridges across the Beatty Cutoff Road to Daylight Pass. The largest one, a former spit, is 1.8 miles northeast of the junction with CA 190, about 9 miles north of the visitor center. It projects 400 to 500 yards eastward from low hills. It is several hundred feet wide at most, round-topped, a maximum of 30 feet high, and steeper on its south, lake-facing side. The beach ridge contains clean, well-sorted, and smoothly worn tabular beach pebbles. Look in roadcuts and you'll see that the pebbles are cross-bedded and shingled. These pebble features are prima facie evidence that the ridge was once the shore of a lake. The top of this beach is 150 feet above sea level, which suggests a maximum lake depth around 430 feet here, ignoring possible fault-related complications. Stand here, look south to the valley, and imagine what the lake would have looked like when it was deep enough to build this beach ridge so high above the valley floor.

Let's be generous and assume a maximum lake depth of 600 feet. The lake's north end, then, would have been about opposite Titus Canyon, and the south end in the broad flat valley north of the Avawatz Mountains, about 100 miles away. The lake's width in central Death Valley ranged between 7 and 8.5 miles, but to the north and south the lake was wider and its shoreline was more irregular.

Any lake 100 miles long and 600 feet deep contains a lot of water. Where did it all come from, considering the aridity of the surrounding country today? The pluvial climate around Death Valley was cooler and

The Beatty Cutoff Road crosses a Lake Manly beach ridge at the double-walled roadcut 1.8 miles north of CA 190. The ridge consists of pebbles that were ground flat by sloshing back and forth in the surf. View is north toward Corkscrew Peak, the prominent peak at right.

wetter during glacial periods, when huge ice sheets covered large parts of North America, than it is today. Although none of the mountains immediately adjacent to Death Valley bore glaciers, they certainly had considerable snow, and the Sierra Nevada, within the expanded Death Valley drainage area, harbored huge glaciers. Even the San Bernardino Mountains had a few small glaciers on their highest peaks during pluvial times.

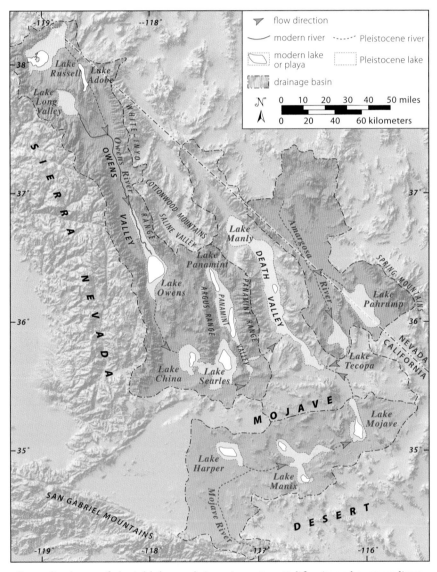

Generalized map of pluvial lakes and rivers in eastern California and surroundings, showing known and suspected connections. Not all these lakes were at peak levels at the same time. At times of peak flow, Lake Manly was fed by the Amargosa, Mojave, and Owens Rivers.

Three large pluvial rivers emptied into Death Valley: the Amargosa, rising from the east in the lofty Spring Mountains of western Nevada (Charleston Peak, 11,916 feet); the Mojave River (Vignette 16) from the south, born among 8,000-foot peaks in the San Bernardino Mountains; and Owens River from the west, draining the ice- and snow-covered 13,000- to 14,000-foot peaks of the eastern Sierra Nevada. Water arrived from the Sierra Nevada after passing through four large pluvial lakes: Owens, China, Searles, and Panamint. Besides runoff from local mountains, groundwater rising to the surface in springs, mostly on the east side of Death Valley and along the Amargosa River, contributed significantly to the nourishment of Lake Manly. This slower and more sustained groundwater flow helped moderate Manly's water-level fluctuations.

Death Valley is one of the best-watered parts of southern California's deserts in spite of its extreme aridity. Large springs, including Nevares, Texas, and Travertine, grace its eastern side, and Travertine Springs sustain Furnace Creek, one of the two perennial streams in the valley. The other perennial stream, Salt Creek (Vignette 11), is maintained by groundwater discharge from the Mesquite Flat drainage area. Much of the meager surface water of the current salt flat comes from groundwater seepage. Large floods of the Amargosa River still flow, albeit infrequently, all the way to Badwater Basin. Today, Mojave River water travels on occasion only as far as Silver Lake playa, about 5 miles north of Baker, and the Owens River has flowed only as far as Owens Lake in historic times.

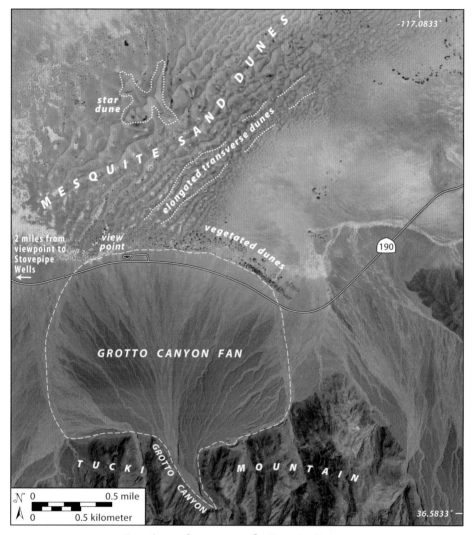

Location and access map for Mesquite Dunes.

GETTING THERE

The winds have piled up a complex of well-formed sand dunes a couple miles east-northeast of Stovepipe Wells. The CA 190 overlook is the best place to enter the dunes on foot, although you can reach them from other nearby spots along the highway. Just travel to or toward Stovepipe Wells on California 190 and stop at the designated dune overlook 1.2 miles east of the village. You can easily reach the edge of the dunes from the overlook by walking 400 feet across the foot of the alluvial fan along well-worn paths. Keep your shoes on until you have walked a ways into the dunes because the sand near the margin contains thorny plant material.

Most people head for the obvious high section, so if you want to see untrammeled sand, strike off for a less obvious destination. The dunes have far more to offer than appears from the CA 190 overlook. Spending just one hour in them can be rewarding, and wandering for several hours is better. If possible pick a cool day, wear a hat and sunblock, and carry water. Allow at least a few hours for a hike to the central high dune and back.

VIGNETTE 9

MESQUITE DUNES
Wind at Play in Nature's Sandbox

Sand dunes fascinate us with their aesthetic purity, their graceful curves and forms, and their highlights and dark shadows in the low-angle light of sunrise or sunset. They record constant activity in small surface markings: animal tracks, ripple marks, branch-swept scars, sand avalanches, and internal layering. Dunes are dynamic, changing shape, size, and sometimes even orientation with major windstorms. They move forward and backward, expanding and shrinking as nature dictates. With every visit, you will find them always intriguing. Grudgingly they house some of the local vegetation, especially around the edges where the sand mantle is thin.

The Mesquite Dunes near Stovepipe Wells are low mounds or ridges of windblown sand on the floor of Mesquite Flat. The sand is largely grains of quartz derived from weathered granite in the Cottonwood Mountains to the west. The quartz grains are transported by floods onto alluvial fans on the east side of those mountains where strong north and northwest winds can blow them out onto Mesquite Flat and eventually into the Mesquite Dunes.

The Mesquite Dunes cover less area and are significantly shorter than many other desert dune fields in California, including Kelso Dunes in the Mojave Desert (*Geology Underfoot in Southern California*) and Eureka Dunes in the remote northern part of Death Valley National Park. The Mesquite Dunes have the same charms as these others, however, and are just a short walk from the highway near Stovepipe Wells. And they are where the robots R2D2 and C3PO crashed in *Star Wars: Episode IV*.

If significant wind is blowing, you'll see that it does not take much of a breeze to move surface sand as long as it is dry. It is curious, then, that the overall dune crests here and in other similar dune fields have been anchored in place for decades. Examination of aerial photographs shows that the dunes have not moved appreciably since as far back as 1982. The pattern was noticeably different around the highest dune in 1948, but in broad terms, the dune field is much the same now as it was more than 70 years ago, even down to the rings of mesquite bushes near the highway.

Walk into the dunes and wander back and forth over their lower parts. Ambitious hikers will probably head for the summit of the highest

Aerial view of Mesquite Dunes looking southeast across the highest dune to the paved highway and toe of the Grotto Canyon fan. This 2006 photo predates the current parking area, which occupies the area between the highway and sand on the right side of the photo.

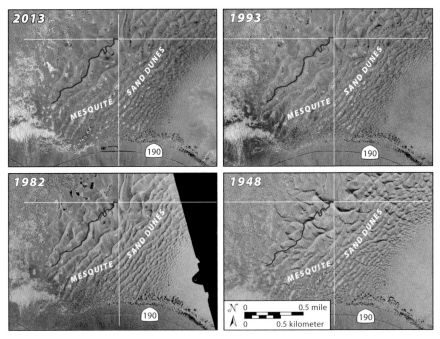

Aerial photographs of Mesquite Dunes going back to 1948. CA 190 arcs across the bottom of each photo, and the parking area is visible in the 2013 photo. A fine white cross marks the position of the highest point in the dunes in 2013, and the wiggly red line marks the crest of the ridge extending southwest from it; these reference marks allow comparisons. The extent of exposed mudflats varies year-to-year as thin sand sheets are blown across them, and in the oldest and newest photos it appears that some of the bushes northwest of the high point were obscured. —Courtesy 2013 Google Earth; 1993, 1982, 1948 US Geological Survey

point, less than 1 mile from the margin. One of the first things to catch your attention on entering the dunes is the size and number of creosote and mesquite bushes that dot peripheral parts of the dunes, especially around the edges where the sand mantle is thin. The bushes are large and relatively lush because the dunes conserve and provide a sustaining supply of moisture. Rain percolates easily through the highly permeable, well-sorted sand, so hardly any runs off. As the surface sand dries, it becomes an insulating blanket, restricting evaporation from the moist underlying sand. Cloudburst floods from the Panamint Range, 1 mile to the south, flow into the dunes and supplement the direct rainfall.

Look for bushes that continue to grow upward in spite of being partially buried in the sand. A clump of mesquite growing out of the top of a dune mound is common, but much of its buried part is deadwood. Wind scouring of the flanks of such a mound exposes these old dead branches.

Smooth, white, mud-cracked silt covers the floors of some dune hollows. Originally, the dried mud formed broad, sinuous bands and lobes within the dunes, but subsequent erosion and burial have separated the bands and lobes into patches. Cloudburst floods intermittently bring muddy water as far as 0.7 mile from Grotto Canyon to form these deposits. Some of the patches are probably remnants of small playas onto which the dunes migrated. Look for photogenic shrinkage cracks that dissect the mud patches.

Dunes come in many shapes and sizes. The simplest form is a ridge transverse to the wind, with a relatively broad, gentle windward slope, inclined around 15 degrees upwind. The Mesquite Dunes are largely an irregular assemblage of mostly short transverse dune ridges. Some long transverse ridges cross the early part of the route from the parking area to the highest dune.

Close-ups of mesquite (Prosopis *species, left) and creosote bush (*Larrea tridentata*, right) at Mesquite Dunes.*

A thin layer of dried mud (lighter-colored areas) overlies sand on the south side of Mesquite Dunes. This mud probably came from a flood out of Grotto Canyon, which cuts the north side of Tucki Mountain in the top center of the photo. The mud layer is dissected by a network of shrinkage cracks.

The highest point in the Mesquite Dunes is the top of a star dune, so called because from above it looks like a marine starfish with several gently curved arms—narrow transverse ridges—converging to a central point. The Mesquite star dune has four converging ridges. For some unknown reason, possibly an unusually dense cluster of mesquite bushes on Mesquite Flat or a confluence of wind currents, the area of the star dune accumulated sand more rapidly than the rest of the dune complex. As that pile of sand grew, it became an obstacle to sand transport, which allowed it to capture more sand and grow still higher. Ultimately, winds began to detour around it and built transverse ridges on its flanks. The star dune's height fully exposes it to the vagaries of multiple wind directions.

Wind transports sand grains in three ways: by rolling or sliding them along the ground; by making them hop; and by impacts of hopping grains as they come back to the ground and hit surface grains, sending them jerkily creeping over the ground. Hopping is by far the most effective and efficient mode of transport for sand, but impact creep enables wind to transport larger particles than it can move either by hopping or rolling. In moderate to high winds, hopping grains form a visible curtain 1 foot or more high, hugging the ground above the dune's surface.

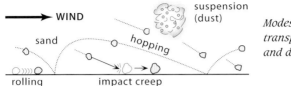

Modes of wind transport for sand and dust.

A dune starts to form as a small, smoothly rounded mound of sand, but strong storm winds eventually create a steeper face on the downwind, or leeward, side. That face becomes so high and steep that the curtain of hopping grains, which normally hugs the surface, separates at the mound's brink, and the sand grains fall onto the leeward face, building it still higher and steeper. At that point, the little sand mound has become an efficient sand trap and a true dune. The lee face continues to grow steeper because sand grains rolling and creeping up the windward slope are dumped onto the upper part of the leeward slope. Hopping sand curtains can form in even moderate breezes if the sand is dry. Watch for these graceful sand veils on the windward sides of dunes.

Dry, loose sand can accumulate at a repose angle of about 34 degrees. Once deposited, dry sand soon slumps, forming a lobe of flowing sand called a sand avalanche. If you explore the dunes shortly after a strong wind, you'll see the scars of sand avalanches on the lee sides of dunes that lay crosswise to the recent wind. You can artificially trigger sand avalanches on freshly shaped lee slopes simply by walking along the brink between the windward and leeward sides of the dune. Try it, it's fun to watch the flowing sand lobes. Big sand avalanches generate a low-pitched moan that resembles the sound of the horn on a distant diesel locomotive or of a propeller-driven airplane. You can sometimes trigger these sounds by starting a large avalanche.

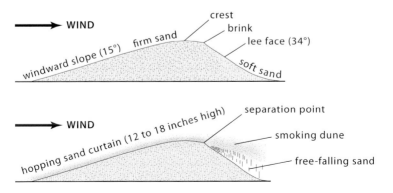

Sand moves up the windward face of the dune, primarily by hopping, and then falls over the crest onto the lee face. The hopping sand curtain crosses a transverse dune.

VIGNETTE 9: MESQUITE DUNES

View is looking at the lee side of a small dune at two sand avalanches with toes. A person wiggled his bare feet in the sand on the crest (top right), creating two small avalanches. Because the tip of each toe produced its own avalanche, the resulting flows have five sand toes each. Note that the wind ripples here are at right angles to the dune crest.

Climbing the lee face of a dune is hard work because avalanche sand is so loosely packed that you slide part way back at each step and sink in up to your ankles. Also, a 34-degree slope is a steep climb, no matter what the material.

An exciting time to be among the dunes is when the wind is blowing hard—30 miles per hour or so. Don't hesitate to enter the dunes during a strong wind but stick mainly to the windward slopes of transverse dune ridges, where the curtain of hopping grains is only 12 to 18 inches high. You will feel the sand on your bare legs, but the rest of you will be comfortable. Avoid the lee face of a dune during a strong wind, for there the hopping grains rain down from the detached sand curtain and get into your eyes, hair, ears—everywhere. The lee face is a thoroughly miserable place to be. By viewing the brink of such a dune in backlit profile, you can see a smoking dune—a curtain of hopping grains separating from the dune surface and taking off into open space. Under strong winds, the brink of a transverse dune can move several feet in just a few hours.

If you want to hike to the top of a sand dune, go up the windward slope. Besides being less steep, the sand is firm there, probably because the curtain of hopping grains that swept across it has compacted it and removed loose grains. Near the brink of the dune, the packed veneer can be thin, and you may break through into the soft underlying avalanche sands of the lee face.

Dunes are covered with sand ripples. The ripples are asymmetrical, like little transverse dunes, with a broad, gentle windward slope and a shorter, steeper leeward face. Most measure 3 to 8 inches from crest to crest—which we call the ripple's wavelength—and the crests rise a fraction of an inch above the troughs. They accurately record the direction of wind currents at the ground level, which may differ considerably from the wind direction only a few feet above.

If the wind is blowing, see if you can find a windward dune slope where ripples are moving. Mark the crests of two or three ripples—toothpicks, nails, or sticks make good markers—and watch how fast they move. It's also fun to rub out the ripples within a patch about 2 feet square and watch them re-form. With a tripod and time-lapse photography, you can make 10-minute videos that show just how much the dune face changes. Ripples that form in windblown sand are sometimes preserved in sandstone rocks millions or billions of years old. These ancient ripples tell us about environmental conditions at the time the sediments were deposited.

Take an early morning excursion into the dunes and examine the tracks of nocturnal prowlers: Coyotes, foxes, and rabbits, as well as smaller inhabitants such as kangaroo rats, lizards, snakes, birds, and beetles leave signs of their travels in the sand—just as the wind has left its mark in nature's sandbox.

Finer ripples superposed on normal-size ripple marks on dunes. Red pocketknife is 3.5 inches long.

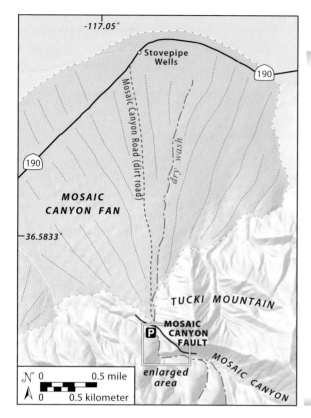

GETTING THERE

Drive 0.15 mile southwest from Stovepipe Wells on CA 190 and turn left onto a wide, well-graded gravel road that leads 2.4 miles south up the bouldery Mosaic Canyon alluvial fan. The road is locally stony and usually washboarded, but passenger cars can easily traverse it. Park at the canyon mouth and walk up the well-worn trail that ascends the flat floor of the wash. To orient yourself, remember that for most of the walk, upstream is nearly due south, so east lies to the left and west to the right. This is definitely a place to visit and photograph more than once, and a place to avoid because of flash floods if it is raining anywhere in the park.

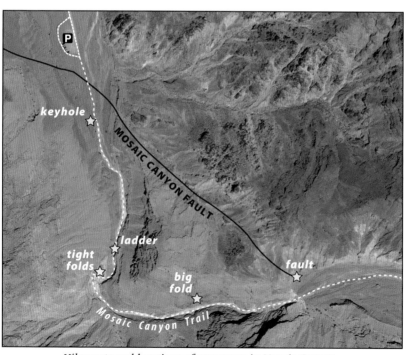

Hike route and locations of structures in Mosaic Canyon.

VIGNETTE 10

MOSAIC CANYON
A Cut-and-Fill Saga of Floods and Debris Flows

Every canyon, every wash, and every alluvial fan in Death Valley was shaped by water. "What water?" you ask. Even though you won't see much water in today's sere landscape, evidence of its power and erosional effectiveness is apparent everywhere in the park and is no more vividly displayed than in Mosaic Canyon, a slot canyon on the north flank of Tucki Mountain near Stovepipe Wells. The easy hike passes through a complicated set of nested debris flows that records multiple episodes of scour and fill over thousands of years. The narrow passage, sometimes filled with gravel debris, is carved into colorful marble that was folded and refolded over millions of years.

The predominant bedrock of the lower canyon is the late Proterozoic Noonday Formation, a seafloor deposit about 600 million years old. Most of the formation consists of tan dolostone, a layered rock formed of the carbonate mineral dolomite ($CaMg(CO_3)_2$), but here most of the Noonday rocks are made of limestone, a rock formed of another carbonate mineral, calcite ($CaCO_3$). To complicate matters further, much of the limestone has been deformed, metamorphosed, and recrystallized to marble. We won't see much of the marble bedrock, however, in the lower parts of the canyon.

Young gravel deposits cover the marble and line the lower canyon walls for the first few hundred yards. Most consist predominantly of layers of pebble- to cobble-size, angular fragments of various carbonate rocks with minor wear. Some layers contain stones of reasonably uniform size—that is, they are well sorted—with platy stones that lie more or less parallel to the layering. The stones are embedded in a sandy matrix. The sorting and orientation of the stones show that these layers were deposited by running water. Closer inspection reveals that many of the platy stones overlap like shingles in a manner that indicates water flowed left to right as you face the west wall of the canyon, the direction water flows in the modern wash.

In contrast, thicker layers in the streambank include larger cobbles and an occasional boulder. Stones in these layers are not as well sorted; platy stones lie at all angles, and only a few touch one another. The matrix contains much fine silt; if it were wet, we would call it muddy.

The mouth of Mosaic Canyon cuts into gravel units that misleadingly appear to be stacked up like pancakes, the way sedimentary rock layers ought to behave. Here, looking upstream, two gray units are exposed on the west side of the wash with a third brown unit on top. The lower unit, which we call the Keyhole unit, forms a bench about 6 feet above the floor of the wash; above it is the Cliff unit.

This cylindrical embayment was eroded into the Keyhole unit, which rises to the level of the person's head, with a vertical cliff in the Cliff unit behind. Note the angular 4-foot boulder embedded in the Keyhole unit (under her hand). More importantly, notice that the Cliff unit lies behind, not on top of, the Keyhole unit. View is of the west wall of the canyon and the stream flows left to right (northward) in this view.

These layers were laid down as a flowing sheet of wet, semifluid, stone-rich, muddy detritus called a debris flow.

The level of gravel in the gorge varies dramatically from year to year, a testament to the power of flash floods that roar through this narrow canyon to scour and fill it. Unraveling this scour-and-fill history would be nearly impossible were it not for a mechanism that quickly cements the deposits in place. They are rich in fine carbonate debris, which percolating water readily dissolves and redeposits, firmly and rapidly cementing the mass of rocks, sand, and gravel. Carbonate cementation may occur within a few years in some places; in Death Valley, it may take a decade or two under the dry conditions of the present day. But thanks to this process, we can work out a history of debris flows and gravel deposition along the walls of Mosaic Canyon.

Let's begin our study of how the canyon and its rocks formed about 100 yards up the canyon from the parking area, where, on the west side, you will come to a vertical, cylindrical embayment, about 4 feet in diameter, that was eroded into the rock unit by water pouring out of a shallow gully above. Let's call this lower rock the Keyhole unit and the cliff-forming rock above it the Cliff unit. Stand back and admire this exposure.

In a normal, depositional, layer-cake succession of sediments, the youngest layer is at the top and oldest at the bottom. However, that is not true here. The Keyhole unit is largely debris-flow deposits with little layering and a large boulder or two. In contrast, rocks exposed at the same height in the back of the cylindrical embayment are well-bedded stream deposits, quite unlike the debris flow deposits of the Keyhole unit. They look just like Cliff unit sediments farther up the wall, and that's what they are. The Keyhole unit was deposited *against* a steep embankment eroded into the Cliff unit, rather than being a continuous layer upon which the Cliff unit was deposited. The same relationships

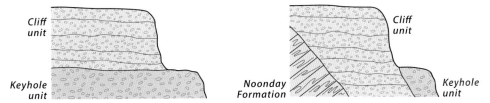

Two interpretations of the west side of the mouth of Mosaic wash. In the first one, the Cliff unit lies upon the Keyhole unit, which is therefore the older of the two. In the second, the Keyhole unit is younger than the Cliff unit and was deposited against it in a gully, just as modern wash sediments are being deposited against the Keyhole unit. The first interpretation seems obvious, but a little detective work will show you that it is incorrect. Furthermore, both units were deposited against tan bedrock of the Noonday Formation.

are under your feet; the modern wash sediments are banked against the steep wall that has been cut into the Keyhole unit, and if they are left undisturbed long enough to be cemented, they will form a yet-younger unit deposited in an incision in the Keyhole unit, which was itself deposited in an incision in the Cliff unit.

Clearly, Mosaic Canyon has been filled and scoured more than once. Farther up the canyon, we will see evidence that such scour-and-fill cycles have been repeated many times. The Cliff unit has been deeply undercut in several places on the west side of the wash. Continued erosion will eventually cause the overhanging roof of these cuts to collapse, dumping well-cemented gravel onto the present streambed. This is but one of several mechanisms that help remove old fills from the canyon walls and thereby widen Mosaic Canyon.

Mosaic Canyon's name comes from stream-polished patches of spectacular sedimentary breccias—accumulations of angular rock fragments embedded in a finer matrix—that are preserved in its walls. Patches of the marble breccia are glued here and there to the bedrock walls of the canyon and are remnants of the oldest canyon filling. The marble breccia forms layers and lenses within well-cemented, gray gravel deposits of mixed carbonate-rock fragments. Examine these layers and lenses

On the west side of the wash, above modern stream gravels at the bottom of the photo, layered bedrock of the Noonday Formation (under book) is overlain by relatively fine brown gravels that grade upward into increasingly coarse breccia that is in turn overlain, at the top of the photo, by more fine brown gravel. All the gravels are well cemented. A thin patch of brownish-gray cemented gravel, a remnant of an earlier canyon fill, is stuck on the left side of the outcrop. Width of view is about 15 feet.

carefully and note that the marble fragments are mostly suspended within a gray sandy matrix. Fragments as large as these, if transported by water, would come to rest touching each other, but they remain separated by the plentiful muddy matrix, indicating that a muddy debris flow, not running water, transported and deposited the breccia.

Some tens of years ago, the steep walls of water-smoothed marble bedrock of the Noonday Formation narrowed into a slot only a few feet across, but the level of gravel in the canyon was high and wide enough in 2021 to completely bury this slot. Amid the attractive sculpturing of bedrock within the narrows, small patches of firmly cemented mixed breccias adhere to the walls. Their configuration can be confusing if you erroneously think of them as extending deep into the wall. Most are merely thin remnants of debris flows plastered onto the wall, more like a Band-Aid than a plug or cavern filling.

In the middle of the narrows you will find a few rebar steps and an iron-pipe railing, the remains of a ladder for hikers to get past a barrier

The much-photographed layered marble, visible to the right of the two rungs of the mysterious ladder in 2016, was completely buried under another two feet of gravel by 2020. Inset from 2003 shows gravel level two feet lower than in 2016, exposing even more of the marble.

that is no longer there. The origins of this ladder—who put it in, when, and why are lost in the mists of time. It may have been used to get over a dry waterfall in the slick marble, but this interpretation is contradicted by the observation that the railing tops out along a near-vertical face about 10 feet above the gravel floor. This would be exceptionally bad planning unless the gravel on the upstream side was once far higher than now. Perhaps a huge boulder once blocked the narrows, backing up gravel behind it. No one seems to know or remember, so the mystery of the ladder remains unresolved to this day.

In the 1980 edition of *Inside Death Valley*, Chuck Gebhardt wrote, "Mosaic Canyon, for example, is nowhere near its former beauty and depth of the early 1970s when iron railings were required for assistance along its precipitous sides. Many heavy rains have washed down tremendous amounts of silt and gravel to fill in the once deep-cut turns in the canyon, completely covering whirlpools that were once so common here."

The year-to-year changes, wrought by flash floods that are rarely witnessed, are remarkable. In 2015 the gravel level was several feet higher than that in 2009, and in 2021 another several feet higher still, completely filling in the slick marble gorge of 2009. Future floods will likely flush out this gravel, repeating the cycle.

(left) Students on a geology field trip descend the narrows in Mosaic Canyon in 2009, with 2015 and 2021 (right) gravel levels superposed. All but the last two people would have been fully submerged in 2015, and only the top of the last student's head would have been above the gravel in 2021.

About 100 feet upstream from the ladder railing, the canyon walls widen to about 20 feet and consist mainly of well-cemented mixed gravel, not bedrock. You previously observed downstream that the layering in such gravels inclines gently downstream. As you move through this less restricted area, however, pay close attention to the east wall. Within 35 feet, you'll encounter two large patches of cemented gravel with near-vertical bedding. How can areas with such steep bedding lie within layers of essentially identical, nearly horizontal deposits? The answer is obvious now that you know the scour-and-fill history of the canyon—the two large blocks fell into the canyon. The entire wall has been deeply scoured into overhangs by floods that must make a hard left turn on their way down the canyon. It is easy to see how blocks of cemented breccia can fall into the canyon and be incorporated into younger deposits.

Consider one final scour-and-fill question: Why do episodes of scouring alternate with periods of filling? On long time scales, climate changes and faulting at the range front will affect the sediment balance in the canyon, but these processes cannot explain changes from year to year. These changes are probably driven by the localized nature of desert rainstorms and the great variation in the amount of loose debris on the

Well-cemented breccia above the narrows has two strange patches with vertical bedding (outlined with dashes) in front of the second author. These blocks fell from an overhanging ledge. Behind him, the breccia has been eroded by flash floods, exposing a vertical wall of horizontally layered Noonday marble.

mountain slopes. A hard rain falling on bare rock will send high-speed floods into the canyon, entraining gravel and moving it downstream, whereas rain falling on slopes covered in loose rock will refill the canyon with debris flows. Furthermore, gravel mobilized in higher parts of the canyon by intense rainfall may move downstream but stop as water drains from the flow, leaving slugs of rock debris that will be moved down the canyon in the next flood.

Now that we have gained some insight into the remarkable story of Mosaic Canyon's long history of scouring and filling, let's pay attention to the bedrock. You have undoubtedly noticed that the marble is layered, and it is reasonable to assume that this layering is sedimentary bedding. However, these rocks have been so intensely folded and faulted that the nature of the layering is obscure. Walk up the canyon another 100 feet or so past the vertical gravel bedding, around a wide 90-degree right-hand bend to broad marble outcrops that face a steep south-facing marble cliff. The marble is slick; walk with care. This is a good place to sit and ponder what you see in the cliff.

The south-facing cliff consists of gently dipping bands that are more or less parallel, but look closely: these bands are folded so intensely that the

Looking north at a face in Noonday marble, a short hike upstream from the exposure of vertical gravel bedding. Although the layering resembles sedimentary layering, it was produced by intense folding. Two folded layers are traced for you; see how many others you can find.

This broad open fold in the Noonday Formation upstream of the narrows contrasts dramatically with the tight, isoclinal folds elsewhere in the canyon.

fold limbs are parallel—they simulate bedding, but if you are standing on one layer and consider its top to be "up", then if you follow it around the fold hinge, you'll be on a limb where that "up" is now "down." The same degree of folding exists in marble lower down the canyon, but it is harder to see. Many bands cannot be traced far before they terminate at a subtle fault, or wrap back on themselves like a piece of ribbon candy. These rocks have been severely deformed, and several lectures in a structural geology course could be based on this outcrop alone.

About 0.25 mile above the bend, the canyon widens dramatically and turns from due east to southeast. At this point, a narrow tributary joins from the south. Watch the canyon floor as you enter this more spacious area. Outcrops of Noonday marble peeking through gravel on the canyon floor end abruptly along a straight line, and the northeast wall of the main canyon beyond and to the left consists of wholly different rocks. You are entering the Mosaic Canyon fault zone, a complex major structure of the Tucki Mountain fault block. The canyon swings to the southeast here because it is following the fault. Like humans, nature likes to take the path of least resistance and therefore erodes the canyon along the highly fractured, and therefore more easily eroded, rocks of the fault zone. Inspect the northeast canyon wall for about 100 feet farther upstream. It consists of about 75 feet of buff, ground-up rock and tectonic breccia, which is the fault zone, and beyond that, of shattered, fine-grained, greenish shaley rocks of the Proterozoic Johnnie

Formation, which is younger than the Noonday Formation. The fault zone continues southeast up the floor of Mosaic Canyon, with much fault breccia on the left (northeast) wall. Tucki Mountain is a remarkably complicated piece of the Earth, but then so is most of Death Valley.

If you continue up the broad wash for about half a mile past the point where it opens up, you will come to a place where the canyon narrows and is blocked by large boulders of cemented gravel. Although it is possible to scramble past this blockage without rock climbing, further passage is blocked by more large boulders and slick marble waterfalls, so this is a good place to turn around and head back. You should consult *Hiking Death Valley* by Michel Digonnet if you wish to explore this little-visited part of the canyon.

These large blocks may someday be incorporated into a younger unit. It is easy to imagine overhanging cliffs upstream of this blockage contributing large boulders of cemented gravels to the boulder pile.

At the narrows upstream of the wide part of Mosaic Canyon the channel is blocked by huge boulders of cemented gravels. Given all the overhanging canyon walls we've seen, this process has probably been common in canyon history.

VIGNETTE 11

SALT CREEK
Posted: No Fishing

Salt Creek is one of only a few perennial streams in Death Valley and one of the few remaining habitats of the Salt Creek pupfish (*Cyprinodon salinus*). These little survivors benefit from a tectonic party going on in their neighborhood, courtesy of the northern Death Valley fault zone. Salt Creek connects Mesquite Basin, home of the Mesquite Dunes (Vignette 9), with Cottonball Basin, the large salt flat north of Furnace Creek. To do so it has cut a gap through the low Salt Creek Hills. These hills are the surface expression of a growing anticline that has brought impermeable clay layers in the subsurface to shallow levels. This forces groundwater to the surface, to the great benefit of pupfish.

A line of low hills parallels the road about a quarter mile northeast of the turnoff to Salt Creek. These hills have been planed off on the side facing the road by the northern Death Valley fault zone, which runs up the west side of the Grapevine Mountains and plays a major role in opening the valley, just as faults along the west side of the Black Mountains do. The Salt Creek fault and Salt Creek anticline are minor

A solitary male pupfish, about 1.5 inches long, in Salt Creek.

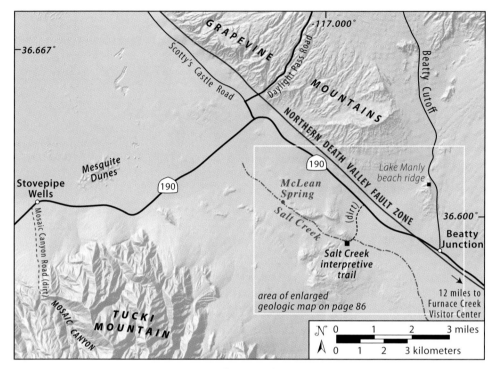

Location of Salt Creek trailhead.

GETTING THERE

The unpaved road to Salt Creek is 13 miles northwest of Furnace Creek along CA 190. The turnoff to Salt Creek is 2.4 miles beyond the cutoff to Beatty. Take this good gravel road for 1.2 miles to its end at the trailhead.

structures related to movement along this fault system. They serve to bring clay-rich beds of the Furnace Creek Formation to the surface, and the surface expression of Salt Creek coincides with these light-colored, impervious beds.

The small, almost pitiable Salt Creek provides an extraordinary home for the rare pupfish in one of the hottest, driest, and harshest environments in the world. The fish is a survivor of the drying and warming climate that followed the last pluvial stage, which ended about 15,000 years ago. During pluvial times Death Valley was filled by Lake Manly (Vignette 8), which was fed by the Amargosa, Mojave, and Owens Rivers. At various times this large system was connected to the huge Lake Lahontan Basin in western Nevada, and perhaps to the Colorado River via the Mojave Desert. Fish moved freely among these basins. Drying and warming of the climate over the past 15,000 years progressively cut these links. Eventually Lake Manly dried up, leaving behind smaller and smaller water bodies, which became more and more isolated until they eventually disappeared.

Today, members of the pupfish genus *Cyprinodon* exist in a handful of isolated environments, and the isolation of these habitats led to the evolution of different species. In the Death Valley region, the Salt Creek pupfish is restricted to Salt Creek; the Cottonball Marsh pupfish to the marsh a few miles south of Salt Creek; the Saratoga Springs pupfish to Saratoga Springs in the southeastern corner of the park; and three members to parts of the Ash Meadows area in western Nevada. These include the Devils Hole pupfish, which lives in the shallowest part of an extensive cave system. Pupfish are of special interest because divergence of these species occurred in a remarkably short time, just the 15,000 years or so since the end of the last pluvial period. All are threatened by development, groundwater pumping, and other aspects of human land use.

The interpretive trail is a flat, 1-mile-long boardwalk that takes you over desert sand and along the creek. Soft, light-colored sedimentary rocks along both sides of the creek are tilted moderately to the northeast. The small pupfish are most visible in these waters in the spring when they are spawning. Even if you don't see the fish, this is still a wonderful trail in Death Valley.

Salt Creek along the boardwalk.

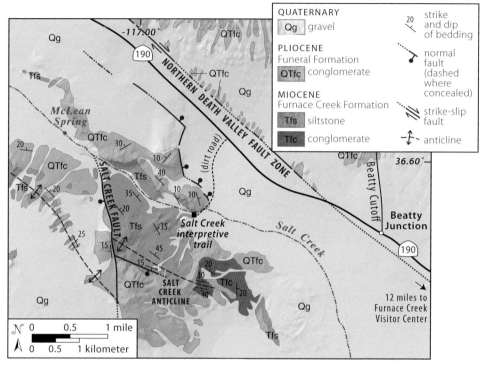

Generalized geologic map of the Salt Creek Hills and Salt Creek. —Wright and Troxel, 1993

Salt Creek originates near Mesquite Dunes and flows south for 25 miles to Badwater Basin, but most of the time only a few stretches have surface flow. A portion of Salt Creek's salty waters emanate from McLean Spring, a few miles northwest of the parking area, and Salt Creek Spring, which is in a gully off the end of the boardwalk. Water from the former spring goes underground immediately but resurfaces when it hits the uplifted claystone of the Furnace Creek Formation in the Salt Creek Hills. The creek normally flows as far as the trailhead between November through May, and then it soaks back into coarse conglomerate beds near the parking area and flows underground to Cottonball Marsh. During the rest of the year, water—and pupfish—are at the far end of the trail.

Pupfish have adapted to survive in water of highly variable salinity. Like other saltwater fishes, their body fluids, with about 1.5 percent by weight of salt (NaCl), are less saline than the water they live in. Consequently, they leak water from their bodies by osmosis into their surroundings and must drink in salty water to replenish it. Their bodies are well adapted to excrete this excess salt via special cells in their gills, but there are limits, and pupfish are rarely found in water more than 2.5 times saltier than seawater. Remarkably, pupfish can survive in distilled water for extended periods of time. Apparently the little pupfish, under the high-stress environment of Death Valley, where water salinity may reach twice that of seawater during a hot summer and then plummet to freshwater levels when the stream is inundated by a thunderstorm, has adapted to these demands in ways that other fish have not.

The pupfish are Death Valley's largest aquatic vertebrate at 1 to 2 inches long. You'll have to look closely to spot these little guys because they dart quickly from place to place. They like to hide under tasty algal mats, under overhanging plants, and around the boardwalk pylons.

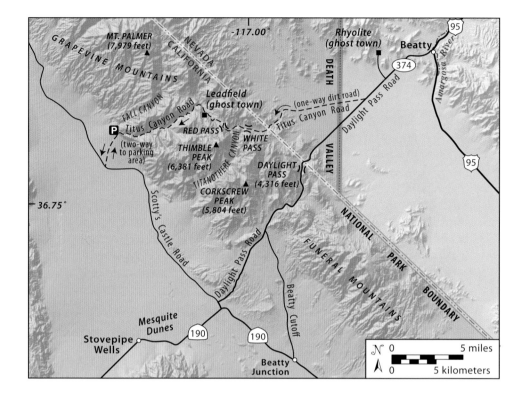

GETTING THERE

The one-way road through Titus Canyon starts in Nevada from the road to Beatty. From Furnace Creek Visitor Center, go 10.7 miles north on CA 190 to Beatty Junction. Turn right onto Beatty Cutoff Road and follow it uphill 10 miles to its intersection with Daylight Pass Road, driving through a large Lake Manly beach ridge (Vignette 8) on the way. Bear right, exit the park, and head for Nevada. The turnoff, marked by a brown National Park sign that reads "Titus Canyon," is 6.6 miles east of the Nevada state line. If you are driving southwest from Beatty, Nevada, on NV 374 toward the park, the junction with Titus Canyon Road is 2.2 miles beyond the intersection of NV 374 and the main road to Rhyolite ghost town.

Do not attempt the Titus Canyon drive if it has been raining because the road can be slippery, and the lower part of the canyon is a slot subject to flash floods. The road is a narrow, bumpy, locally steep, one-way dirt road of 28 miles. It is accessible to 2WD vehicles with good clearance when graded, but floods often wash it out, so check current road conditions at the Furnace Creek Visitor Center or at the park web site. Take food and water and a good spare tire; the nearest supplies and gasoline are in Stovepipe Wells and Beatty, and as in much of the park, there is no cell phone service.

If you do not want to commit to a full-day excursion, the spectacular lower narrows and its unusual breccia can be reached by hiking in from below. The two-way road to the canyon mouth leaves the Scotty's Castle Road 14.9 miles north of its junction with CA 190. Drive 2.6 miles up the gravel road to an ample parking area. The breccia is an easy 0.7 mile walk up the narrow canyon from there.

VIGNETTE 12

TITUS CANYON
A Geologic Crime Scene Turned Upside Down

The 28-mile drive over the southern Grapevine Mountains and down Titus Canyon is always a highlight of a trip to Death Valley. With awe-inspiring scenery and spectacular geology, the drive presents plenty of opportunities to put boots on the ground and see interesting geologic features up close.

First, a warning. The geology of the Grapevine Mountains is complicated beyond description and only partially understood. Like many geologically interesting areas, the rocks in this range have been successively stretched and ripped, squeezed, folded, torn, and cooked; dismembered again; flooded by rivers; stomped on by great prehistoric beasts; and scalded by hot lava and tephra. The Titus Canyon drive goes right through this geologic crime scene, where great sections of the sedimentary rocks have been flipped upside down. We will do our best to keep you upright.

Three primary sequences of rocks are exposed in Titus Canyon. From oldest to youngest, these are late Proterozoic and Cambrian marine sedimentary rocks; Eocene, Oligocene, and Miocene mudstone, sandstone,

Rocks in upper, eastern Titus Canyon consist largely of colorful volcanic and sedimentary rocks of Cenozoic age. Here pinkish Miocene ash-flow tuffs and brown lava flows on the left skyline, about 13.8 to 11.5 million years old, overlie coarse red and green conglomerates of the Titus Canyon Formation, about 38 to 30 million years old. Although cut by numerous faults, these rocks are upright.

Rocks in the lower, western part of Titus Canyon are largely upside-down Paleozoic rocks. Although layered, they are significantly less colorful than the Cenozoic rocks to the east. Here at the narrow exit of Titus Canyon, in the steep western face of the Grapevine Mountains, reddish-brown beds in the foreground are upside-down Bonanza King Formation; thinly bedded light- and dark-gray layers in upper right are steeply tilted beds of Bonanza King Formation; and gray beds in upper left are upright Bonanza King overlain by younger Paleozoic sedimentary rocks culminating in Silurian rocks on Mt. Palmer.

	Age	Formation	Graphic Column	Rock Character	Thickness (feet)
CENOZOIC	PLIOCENE AND MIOCENE	*unconformity*		volcanic rocks, tuff	
	MIOCENE	Panuga *unconformity*		conglomerate, sandstone, tuff	50-200
	MIDDLE EOCENE TO EARLY OLIGOCENE	Titus Canyon *unconformity*		river conglomerate, mudstone, red beds	1,850
PALEOZOIC	CAMBRIAN	Bonanza King		dolostone, limestone	3,500
		Carrara		shale, siltstone, limestone	1,530
		Zabriskie		quartzite	950
PROTEROZOIC		Wood Canyon		quartzite, siltstone, dolostone	3,400
		Stirling		quartzite	2,000
		Johnnie		mixed sedimentary shale, dolostone, quartzite, limestone, conglomerate	4,000
		Noonday *unconformity*		dolostone, limestone	1,000

Generalized stratigraphy of rock formations in northern Panamint and southern Grapevine Mountains.

RHYOLITE GHOST TOWN

The legendary prospector Frank (Shorty) Harris found some green rocks resembling a bullfrog speckled with gold on August 4, 1904. His discovery led to the fabulously rich but short-lived Bullfrog Mine and fueled the boom of the town that was Rhyolite. At its peak in 1907, Rhyolite's population of about four thousand relied mainly on gold production from the nearby Montgomery-Shoshone Mine. The town eventually sported three competing railroads, two banks, concrete sidewalks, electric streetlights, fifty saloons, two newspapers, three water systems, two electric plants, an opera house, and a stock exchange. The boom burst by 1911, and the population moved on, leaving behind evidence in concrete of its previous existence.

Disseminated gold in the hill on the southeast edge of town was mined profitably from an open pit from 1988 to 1999, producing nearly 700,000 ounces of gold. The ore was processed on-site. The great tailings basin south of NV 374 attests to that venture.

Drive the main street through Rhyolite to see a few remnants of the town. Look for a house made of 50,000 discarded beer and liquor bottles, the Las Vegas & Tonopah railroad station, and the ruins of a three-story bank.

Only a few decades ago, it was possible to climb around on the Bullfrog Mine's waste dump and find rare pebble-sized green rock with a fleck or two of gold. No longer. When the price of gold rose dramatically about thirty years ago, the waste dumps of many gold mines and prospects in the western US were shoveled into trucks and hauled away to process them for any remaining gold.

Some of the slowly crumbling ruins of Rhyolite.

and conglomerate that were deposited on land; and Miocene volcanic rocks that are mostly tuffs erupted from calderas in Nevada.

To help keep track of where you are, we have broken the trip into four segments marked by recognizable topographic features, with an odometer reset at the start of each leg.

Leg 1: Titus Canyon turnoff to White Pass

Reset your odometer at the intersection of NV 374.

Titus Canyon Road is a sharp westward turn off eastbound NV 374. The road brings you back into the park after 1.8 miles, and it steadily climbs the alluvial slopes of the Grapevine Mountains for about 6 miles before entering a broad wash and continuing to climb. These low hills are mostly rhyolite lava flows and welded tuff erupted 13.5 to 11.5 million years ago in the Miocene Epoch. After about 1 mile, you are back in California. About 3.5 miles from the entrance to the wash, the road tops out at a broad saddle, recognizable because the road starts a long descent on its western side. This low saddle is known informally as White Pass (9.6 miles on the odometer and elevation 5,115 feet).

White Pass is a good place to stop and contemplate the geologic landscape. Looking west, the peaks on the right of the road are more Miocene volcanic rocks, and those on the left are Paleozoic sedimentary rocks that we'll see plenty of farther along the road. The Titus Canyon Formation occupies the low area in between and makes up much of the

View west from White Pass. The layered brown rocks on the skyline to the left of center are Cambrian sedimentary rocks, which are in fault contact with the much younger, light-tan Cenozoic volcanic and sedimentary rocks to the right of center. Dark hills in the left middle ground are megabreccia blocks of dolostone. Thimble Peak is the distant pointed summit left of center.

ground we will drive upon in the next few miles. The formation consists of green, brown, and red conglomerate, sandstone, and mudstone deposited during Eocene and Oligocene time, about 38 to 30 million years ago. The age is noteworthy because rocks of Oligocene age are very rare in this part of California.

White Pass is a drainage divide. A raindrop falling onto its east side could theoretically make it down to the Amargosa River, which flows through Beatty, thence to Death Valley via its southern end, and eventually north to Badwater Basin. A drop falling on its west side would theoretically go down the canyon ahead, the east fork of Titanothere Canyon, thence to Salt Creek, and south past Furnace Creek to end up at the same place.

Leg 2: White Pass to Red Pass

Reset your odometer at the top of White Pass

In about half a mile, the road crosses several shallow washes that lead into the east branch of Titanothere Canyon (look left). About 1.8 miles from the pass, the road is cut into reddish conglomerate and thin sandstone and mudstone layers of the Titus Canyon Formation, the unit that contains the land mammal fossil for which Titanothere Canyon is named. Within a quarter mile, the road passes among lumpy dark-gray outcrops of dolostone from the Bonanza King Formation. These odd outcrops are huge blocks that fell into the basin in which sediments were accumulating, presumably from fault-scarp cliffs that bounded the basin. They are common in this area but are also found in Miocene lake sediments with the Resting Spring Pass Tuff (Vignette 20). Huge blocks such as these are typically shattered, and their deposits are known as megabreccia.

About 1.5 miles from White Pass, the road makes a sharp left-hand hairpin bend and crosses the main branch of Titanothere Canyon. After crossing the wash, the road passes through well-exposed roadcuts of green-brown Titus Canyon Formation conglomerate. These beds underlie much of the road from White Pass to Red Pass.

As noted in the initial 1935 report by Chester Stock and Francis Bode, paleontologists at Caltech:

> Credit for the discovery of vertebrate fossils in the Tertiary series of the Grapevine and Funeral Mountains goes to H. Donald Curry, . . . Acting Ranger-Naturalist at the Death Valley National Monument. During the fall of 1934, Mr. Curry, while traveling over the little-used road between Rhyolite, Nevada, and the now deserted mining camp of Leadfield in Titus Canyon, Grapevine Mountains, saw and collected, in an outcrop of maroon or chocolate-colored, calcareous [calcite-cemented] mudstone exposed in the face of a road-cut, a skull and lower jaws with teeth of a titanothere.

94 VIGNETTE 12: TITUS CANYON

Roadcut exposure of a dark gray megabreccia block made of Bonanza King Formation dolostone on top of Titus Canyon Formation conglomerate in Titanothere Canyon (36.829083, -117.019512). The conglomerate consists of well-rounded pebbles and cobbles, the largest of which are about 5 inches across. Bedding is difficult to see, but interbedded sandstone layers indicate that these rocks have been steeply tilted.

The skull found by Curry was later identified as belonging to the titanotheres and named *Protitanops curryi*. Titanotheres of many sizes roamed an extensive savannah landscape of lakes, lush vegetation, and low mountains in Oligocene time. Other vertebrate fossils recovered from the Titus Canyon Formation include three-toed, 2-foot-tall horses, primitive tapirs, three types of rhinoceroses, bear-dogs, hyena-like carnivores, several types of squirrel-like rodents, camel-like and tiny deer-like herbivores, and two types of unusual pig-like animals, called oreodonts, that have four-toed hooves. All eighteen known species that make up the Titus Canyon fauna are long extinct and most have no close living relatives. Nevertheless, the fauna provides information about what this area was like 38 to 30 million years ago.

Artist's rendition of the titanothere Protitanops curryi. *These rhinoceros-like beasts were 8 feet tall at the shoulder.* —Reproduced here with permission and courtesy of the artist, Nobu Tamura

The road climbs steeply up to its summit at Red Pass (12.3 miles, elevation 5,266 feet), where you may stop and enjoy the sublime views. As at White Pass, the rocks on the right (north) are Miocene volcanic rocks, those on the left are Paleozoic sedimentary rocks; here in Red Pass and in the valley between the two, the red sandstone and conglomerate beds are Titus Canyon Formation almost all the way to Leadfield.

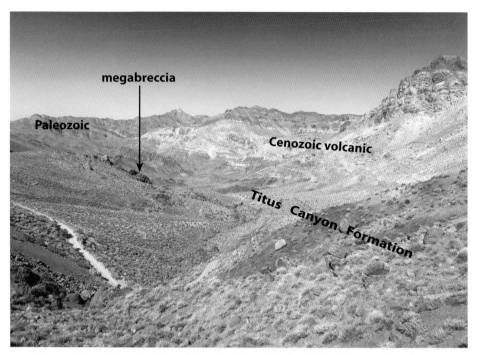

View northwest from Red Pass into colorful east fork of Titus Canyon. From here to Leadfield, about 3 miles ahead, the contact between Cenozoic sedimentary and volcanic rocks on the right and Paleozoic sedimentary rocks on the left is a fault. The distant striped mountains on the left of prominent pyramidal Mt. Palmer are Cambrian Bonanza King Formation carbonates. Red and green rocks in the low ground, below the pink cliffs, are mostly Titus Canyon Formation. The light-colored rocks above the pale-green ones in the center of the photo are Cenozoic volcanic rocks.

Leg 3: Red Pass to Leadfield

Reset your odometer at the top of Red Pass.

The road descends steeply 1,000 feet to Leadfield (3 miles), where Titus Canyon Formation conglomerates are faulted against the steeply dipping Bonanza King Formation that forms the ridge south of the Leadfield site. Like so many other mining ventures and boomtowns in the Death Valley region, this was solely a fraudulent paper scheme promoted by one C. C. Julian in 1926. He floated brochures all over the eastern United States about his Western Mining Company, depicting a beautiful landscape on the shore of a big lake with snowcapped mountains on the near horizon. He built a road up from the valley to bring one thousand potential investors in more than one hundred cars to see Leadfield, his new town. For a brief time, it boasted a post office, telephone, and a newspaper—*The Leadfield Chronicle*. Three hundred hopeful prospectors stuck it out for about a year before concluding that the scanty lead ore was worthless and they had been thoroughly duped, hoodwinked, and swindled. They left behind a few tunnels, pits, mine tailings, and ramshackle buildings. Most of the building foundations are on the south side of the wash, whereas tent sites and retaining walls are scattered all over the canyon floor. Be cautious if you explore, and stay out of the many open pits and tunnels in this area; all are dangerous. Snakes and bats seek shelter in them, rotted timbers might give way underfoot, and loose rock might fall from the ceiling.

The ghost town is a good place to stretch your legs and examine nearby exposures of the Titus Canyon Formation. The conglomerate is composed of well-rounded, smooth pebbles and cobbles of fine-grained sedimentary rocks, including chert, which is fine-grained silica deposited in deep-marine water, and quartzite (dense, well-cemented sandstone, commonly metamorphosed). The pebbles and cobbles were likely brought here from central Nevada by a west-flowing river, one of a number that carried sediment from Nevada to the Pacific Ocean before the Sierra Nevada existed. The former river valley channeled large ash-flow tuffs that erupted in Nevada; some ash-flow tuffs swept down-canyon more than 150 miles from their source calderas.

Bonanza King dolostone beds, folded into a broad anticline, in fault contact with Titus Canyon Formation in the south wall of Titus Canyon above Leadfield.

Titus Canyon beds are steeply tilted around Leadfield, but outcrops such as this allow you to tell which way was stratigraphically up. The finger is on sandstone displaying faint layering; the conglomerate has eroded into the sandstone, cutting through the layering. This demonstrates that the conglomerate is younger than the sandstone, and that stratigraphic up is about 45 degrees clockwise from the direction the finger is pointing. Pebbles and cobbles are composed primarily of black chert and quartzite.

Leg 4: Leadfield to canyon exit

Reset your odometer at Leadfield.

The road makes a hairpin left bend 0.4 mile below Leadfield and enters the first of two pronounced narrows carved in Bonanza King dolostone. Stop about 100 yards before entering the first narrows, when you reach the view in the next photo, and examine the scene ahead. Carbonate beds of the gray-striped Bonanza King Formation, here upright, appear to be folded into a U shape, a syncline. A National Park sign once here entitled "When Rocks Bend" pointed out this fold—except that it is not a fold. The beds are actually planar, dipping to the north, but their intersection with the bend in the wash gives the false appearance of a fold, an illusion that has fooled many a viewer, geologists included.

Up to this point you have been driving in a tributary (east fork) of Titus Canyon. A quarter mile past the false fold, the road meets the main branch of Titus Canyon entering from the right. For about 0.75 mile, the canyon is carved into cliffs of the Bonanza King Formation. The beds in this stretch are tilted but still upright.

The striped Bonanza King Formation beds that appear to be bent into a U shape are actually all dipping about 40 degrees to the north-northeast. This illusion of a fold is created by the intersection of dipping beds with the indentation in the canyon wall.

At Klare Spring, the only reliable spring in the canyon at 2.1 miles downstream from Leadfield, the road crosses the Titus Canyon fault, a major flat-lying fault zone that separates the Bonanza King Formation from a slightly older sequence of sandy and shaley rocks that were deposited in shallow water and near the shoreline. They include, from oldest to youngest, the Wood Canyon Formation, Zabriskie Quartzite, and Carrara Formation. The Wood Canyon Formation here and Deep Spring Formation in the White Mountains are of great interest to stratigraphers, paleontologists, and evolutionary biologists because they contain records of the Proterozoic-Cambrian transition, the time when fossils became abundant in the geologic record. The Wood Canyon and Carrara Formations are more slope formers than cliff formers, and the canyon widens significantly here as a result.

From the Titus Canyon fault at Klare Spring and for the next 2.5 miles downcanyon, the sedimentary layers so elegantly exposed in the north wall of the canyon are flipped over; they dip steeply to the east, but their stratigraphic tops are to the west. This is clear because below the spring you encounter the units in reverse order: first the Wood Canyon Formation, then the Zabriskie Quartzite, and then the Carrara Formation. The Zabriskie contains sedimentary structures that indicate which way was stratigraphically up when those sands were deposited, and they agree with the sequence of formations—up is now upside down. Check out these attractive purplish-pink, overturned Zabriskie quartzite beds at mile 3.8 to see if you can discover some of the sedimentary structures, such as cross-bedding, graded bedding, and channels.

A fine exposure of ripple marks is exposed on a bedding surface at 4.0 miles at a 90-degree right bend in the road on the south side of the canyon. In about half a mile, the road reenters the gray Bonanza King Formation.

About 6 miles downcanyon from Leadfield, we enter the lower Titus Canyon narrows—4 miles long—with towering cliffs hundreds of feet high and locally just 26 feet apart at wheel height. The overturned Bonanza King beds generally look flat but numerous faults and folds attest to severe deformation, which is not unexpected in rocks that have endured a half-billion years.

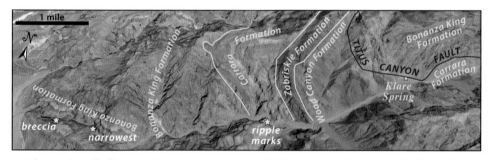

The north wall of lower Titus Canyon exposes a thick sequence of overturned beds. In this Google Earth image, looking down to the north, the Titus Canyon fault (in red) marks the boundary between right-side-up Bonanza King Formation beds, in the upper right, and an upside-down sequence of Wood Canyon, Zabriskie, Carrara, and Bonanza King rock layers. Numerous faults, omitted here, cut all these formations. Labels are oriented to show right side up from upside down. The view extends to the canyon mouth (left).

Canyon-level view of overturned Zabriskie Quartzite in the pinkish cliff-forming layers below the peak, with the varicolored Carrara Formation and dark-gray Bonanza King Formation to the left, also overturned. (36.8273, -117.1112)

Entrance to the Titus Canyon narrows in the typical layered dolostone of the Bonanza King Formation—typical, that is, except for being upside down.

About 1 mile downcanyon from the entrance to the lower narrows (mile 7.0), the canyon walls expose a remarkable mosaic of dark, angular dolostone blocks cemented by white calcite. In some places angular blocks clearly fit back together; in others, large blocks seem to have exploded into swarms of smaller blocks along the edges. These outcrops have puzzled geologists for decades. One reasonable hypothesis is that this volume of rock formed by collapse of a cavern, time unknown, but that explains only some of the features. Faulting along the west front of the Grapevine Mountains may have played a role.

As you continue down the canyon to its mouth, you'll see spectacular evidence for the abrasive action of boulders, cobbles, and sand that have scratched and polished the canyon walls during big flash floods. The occasional coating of mud on the canyon walls reveals the depth of the last major flood through the mouth of the canyon. The canyon walls are undercut in places and will collapse someday, blocking the canyon until water cuts a new channel through the debris.

The road follows the deep and dark canyon downstream from the breccia through several turns before widening a bit and suddenly coming to the canyon mouth in northern Death Valley. Titus Canyon streams and flash floods debouch their sediment onto a huge alluvial fan. Dense debris flows flushed the big boulders that you see out onto the fan.

Parking is available at the head of the fan. A trail leads north from the parking lot to Fall Canyon, another narrow slot, and eventually to the mouth of Red Wall Canyon. A hike up Fall Canyon, over the main range crest north of Mt. Palmer and back down Red Wall Canyon, is an extraordinary trek, but it should be undertaken only when well prepared and with a start early in the day.

The drive from the parking lot to pavement is nearly 2.5 miles down a dusty, well-maintained, two-way gravel road. One tenth of a mile from the pavement, the gravel road climbs over a steep berm that is an east-facing fault scarp in the northern Death Valley fault zone, which extends 100 miles northward from Furnace Creek to Fish Lake Valley. Geologists estimate that the last major earthquake on this fault strand happened about 2,000 years ago, judging by the rounded, eroded nature of the scarp.

Flood-smoothed exposure of cryptic breccia in the south wall of the narrows in lower Titus Canyon. Someone should figure out how it formed.

VIGNETTE 13

UBEHEBE CRATER
A Steamy Story of Magma and Water

The main natural attraction in the remote country north of Stovepipe Wells is Ubehebe (YOU-beh-HEE-bee) Crater, whose walls beautifully display a story of explosive volcanism. Explore this cratered landscape to get a glimpse of violent events that happened a few millennia ago. The crater is one of a set of volcanic explosion craters at the northern tip of the Cottonwood Mountains. Ubehebe gapes nearly 0.5 mile in diameter and drops 500 feet below the elevation of the parking area and 780 feet below the highest point on its east rim. The area around the crater is covered in dark-gray basaltic cinders, but the crater interior exposes older, brightly colored sedimentary rocks that glow orange in afternoon light.

Ubehebe Crater formed about 2,100 years ago when rising magma encountered groundwater, which boiled violently and produced powerful steam explosions that blew rock debris into the air and over the surrounding countryside. Ubehebe is the youngest and largest of around a dozen vents in the area. Unlike cinder cones, which grow from tephra (fragmental volcanic material) that is blown out of the vent and forms a conical pile around it, craters are depressions produced when material is blasted out of the ground. Although Ubehebe has a low rim, it is clearly a crater because most of the rock debris exposed in it is sandstone and conglomerate, not basaltic tephra. The Inyo Craters (Vignette 32) are smaller examples of explosion craters produced by pasty rhyolite magma rather than more fluid basaltic magma.

The debris that settled on the landscape after the Ubehebe eruption contains a mixture of basaltic fragments and the bedrock through which the magma passed. Hillslopes surrounding the Ubehebe area are underlain by relatively soft sedimentary beds which, before the Ubehebe eruption, were intimately dissected by rills and small gullies, 15 to 20 feet deep. The fallout debris drapes spectacularly over this dissected landscape, arching up over ridges and bowing down into and across gullies. Erosion along the gullies has removed much of the fallout debris in them, but fallout draped over ridges remains intact.

The Ubehebe volcanic field contains about a dozen recognizable explosion pits, plus degraded remnants of possibly three or four more, all smaller and older than Ubehebe. These satellite pits cluster in two groups, one a quarter mile west of Ubehebe Crater and the other abutting its south side. The southern group includes Little Hebe, the most distinctive of the subsidiary craters. Near-horizontal lava flows cap tilted sedimentary beds in the low hills a mile and more north of Ubehebe. We know they are older than the craters because they are mantled by debris ejected from the craters.

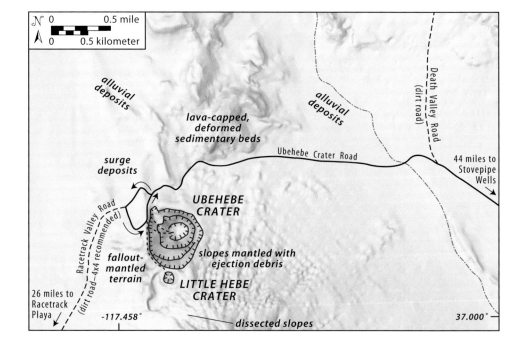

GETTING THERE

Ubehebe Crater lies in the north-central part of Death Valley National Park, about 6 miles west of Scotty's Castle. You can access the crater by a paved road extending 40 miles northwest from CA 190 at a junction 7.5 miles east of Stovepipe Wells. A large parking lot and viewing area are on the west edge of Ubehebe Crater (37.0109, -117.4549). Looking at the crater from the parking area, north is to the left and south to the right. The highest point on the opposite side of the crater is nearly due east, and Little Hebe Crater is almost due south of Ubehebe. From the parking area, you can view the crater from its nearly straight western edge. Good views from other places along the crater-rim trail reveal additional relationships in the crater walls.

Looking north across Ubehebe Crater to the northern Grapevine Mountains. Vehicles at the parking area in the upper left corner of the photo give a sense of scale. The area around Ubehebe Crater was buried under more than 100 feet of dark basaltic tephra, whereas the interior of the crater is largely older sandstone and siltstone, unrelated to volcanic activity. Several steep faults cut the brightly colored sedimentary rocks but not the overlying dark tephra layers, indicating that faulting occurred before the eruptions. In the center of the photo a steep fault places yellowish rocks on the left against orange ones on the right.

Dark, bubble-rich pieces of basaltic tephra formed from frothy basaltic magma blasted out of Ubehebe Crater. Lighter-colored gray, orange, and brown pebbles were torn from underlying sedimentary beds during eruption and went along for the ride.

The ridge encircling Ubehebe consists of thinly bedded, unconsolidated, fragmental volcanic and bedrock debris that erupted out of the crater. Even from a distance you can easily count fifty layers in the encircling ridge, and a close-up count would probably exceed one hundred. Ubehebe must have puffed like a steam engine belching debris. These well-defined layers slope mostly outward but incline gently inward in a few places. The older layered sedimentary rocks in the crater walls tilt about 35 degrees. You can best see this discordance between the deformed and tilted sedimentary rocks and the gently dipping volcanic rock layers on the north and southwest walls of the crater. The sedimentary rocks include beds of stream-deposited conglomerate containing smooth, well-rounded pebbles and cobbles of quartzite, limestone, mudstone, pre-Ubehebe lavas, and granitic rocks. These beds were deposited in Miocene time.

A near-vertical fault cuts through the sedimentary rocks at the crater's northeast corner and separates beds vertically at least 400 feet. The beds of ejected debris drape over this fault, bending down about 30 feet where they cross the fault's surface trace, but they do not appear to be offset by it, indicating that the fault ceased activity more than 2,100 years ago.

In the Ubehebe volcanic field, younger craters cut across older pits, and ejected debris of younger pits mantles older features. So, the craters

The pile of tephra layers on the rim of Ubehebe Crater, seen here south of the parking area, is up to 150 feet thick. Gentle waves in the layers result from the tephra draping topography rather than from deformation.

Aerial view northeast across Ubehebe Crater (center), Little Hebe Crater, and other craters of the southern group. White areas are secondary accumulations of clay and silty pond deposits in the crater depressions. Note intricately rilled slopes in foreground.

must have formed at different times; all of them, however, appear to be the product of a single, relatively short episode of volcanism. Ubehebe Crater is the youngest, and its extensive sheet of debris at least partly mantles almost everything in the volcanic field.

Some of the finer-grained ejecta from Ubehebe was blown out in a devastating base surge. These donut-shaped clouds of gas and rock debris, blasted outward in all directions from the crater, hugged the ground surface, depositing fine-grained rock fragments as they passed at speeds more than 100 miles per hour. A base surge deposits its load by plastering the fast-moving debris onto the crater-facing side of obstacles in its path.

Three trails in the Ubehebe area merit particular mention. The trip down into the crater is easy, but expect a slow, laborious climb out. Don't attempt it on a hot day unless you are in good physical shape. You can also circumnavigate the rim along a 1.5-mile trail to get good views of the crater walls. The most geologically interesting trail is the wide, well-trodden track leading 0.5 mile counterclockwise from the parking area to Little Hebe, which tells a neat geologic story about the succession of events.

Little Hebe Crater (37.0042, -117.4519) is a short walk south of Ubehebe Crater.

A deep gully eroded into layers of tephra southeast of Little Hebe Crater. The tephra layers fell on a landscape that was already gullied, as seen by the way they drape the side of the gully, and then were deeply incised by further erosion.

You'll approach Little Hebe on its north side, but work your way around to the prominent saddle in dark rock on the south rim. By the time you get there, you will realize that Little Hebe is a cone with a large central crater about 300 feet in diameter. In its northwest wall, a thick section of thinly layered tephra from Ubehebe Crater rests upon 5 to 15 feet of dark, volcanic spatter lava. This volcanic spatter arrived at the surface as still-molten lava that sputtered out of a central vent in small molten globs that fell to the ground. Many globs were soft enough to stick together and form a cone of spatter, which Little Hebe's crater-forming explosion later partly blew up. The layer of spatter lava in the walls of Little Hebe Crater is a remnant of the lower slopes of this cone.

A particularly large piece of tephra, a volcanic bomb, ejected from Little Hebe. The erupted lava was clearly molten when it was ejected because its surface was shaped by high-velocity passage through the air. Red pocketknife for scale.

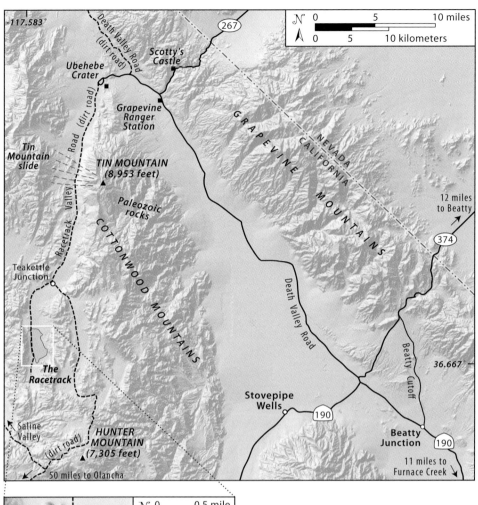
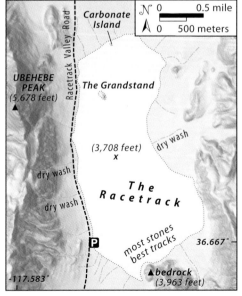

VIGNETTE 14

THE RACETRACK
A Chilling Mystery of Slithering Stones

The sliding stones on the Racetrack, a remote dry lake in the northwest part of Death Valley National Park, have fascinated and flummoxed scientists and visitors for more than seventy years. Hundreds of angular stones, ranging from small cobbles to boulders a few feet across, have left clear trails of their passing in the mud of the playa surface. How, why, and when do these stones move?

Most playas are situated at the toes of alluvial fans, away from steep mountain fronts, and as a consequence their surfaces rarely have any rocks on them. Meteorite hunters know that a lone rock out in the middle of a playa is possibly one that fell from above. The Racetrack differs from most playas, however, in having a steep cliff of dolostone at its southern margin that sheds rocks onto the playa surface. The Racetrack is not the only playa with such trails—they have been reported on at least eight other desert playas in western states—but the abundance of stone tracks on this playa's surface is noteworthy and spectacular.

The Racetrack gained fame about 1950, soon after James F. McAllister of the US Geological Survey presented a report about its rock trails at a meeting of the Geological Society of America. Scientific study and articles

GETTING THERE

The 30-mile-long dirt road to the Racetrack begins from the west side of the Ubehebe parking loop. The unusual stone tracks at the south end of the playa warrant the effort, but the road chews up and spits out tires at an alarming rate owing to sharp blocks of dolostone; drive slowly and watch out for these little tire-biters. The road is suitable for 2WD cars with reasonable clearance, but reread the precautions in the Preface before you begin.

Once you reach Racetrack Valley, go to the far (south) end of the playa where a parking area has been cleared at the playa's southwestern corner. Walk out onto the playa from there (driving on the playa is strictly forbidden and devastating to study of the stones). Stones and tracks are most numerous where the steep mountain face comes down to the playa and extend over an area 0.5 mile north from the parking area. A hike out to the Grandstand and its unusual granite with huge feldspar crystals is also worthwhile.

Please note that Park Service regulations prohibit picking up or disturbing the stones on Racetrack playa in any way.

in popular magazines followed in short order. A photo essay in the 1952 edition of *Life* magazine stated that "Local theories blame this bewitched behavior on the lakebed's tipping back and forth, or on floodwaters, or on Russian tampering with the magnetic pole. Scientists disprove these ideas but have no sure answer of their own." Within a few years two competing hypotheses emerged: one, that strong winds pushed the stones on

This block of dolostone traveled several hundred feet to its present position from the base of the hill on the left edge of the photo.

a wet, slippery playa surface; another, that the stones were embedded in an ice sheet that the wind pushed across the playa.

In 1953 John Shelton of Pomona College landed an airplane on the playa, doused a small area with water, and tried to move stones with artificial wind generated by the airplane's propeller. He reported only partial success. George Stanley of Fresno State College (now California State University, Fresno) surveyed trails in 1955, found that many of them were parallel, and proposed that they had moved in concert while trapped in an ice sheet. Other studies looked at the physics of movement by wind alone and concluded that hurricane-like velocities are required.

Parallelism of trails seemed to require something to bind the rocks together, so Bob Sharp of Caltech (your third author) and Dwight Carey of UCLA tested the ice hypothesis in 1976 by building a corral of iron stakes around two stones. During the course of the experiments, which covered eight years, one moved out of the corral whereas the other stayed put. Based on this and other evidence, they concluded that wind is the primary mover. J. B. Reid and students from Hampshire College concluded in 1995 that large ice sheets and light wind could move the rocks. The two hypotheses remained in competition, and everyone assumed that the conditions under which the rocks moved were so extreme that no one would be able to observe the stones in motion. With this background, let's go to the Racetrack and see what all the fuss is about before we delve back into this remarkable conundrum.

The rock corral experiment of Sharp and Carey. The corral originally surrounded a single stone, which moved out of the corral and left a trail during the winter of 1968–1969. They placed two more stones (seen here) in the corral, and during the winter of 1973–1974 one moved out and the other stayed put. This seemed to indicate that ice was not involved in the movement because it would have frozen onto the stakes and not moved.

VIGNETTE 14: THE RACETRACK

The drive to the Racetrack is long and bumpy, but it covers some outstanding desert scenery. After leaving the pavement of the Ubehebe one-way loop, the road goes through a landscape covered with basaltic tephra for about 1.5 miles and then traverses alluvial slopes and channels made of dolostone and limestone. This stretch of road has claimed an unusually large number of tires, so take it slow and avoid rocks that are sticking up. Ahead, the road goes over the pass west of Tin Mountain. After about 10 miles of gentle climbing, the road passes through low hills and double-walled roadcuts in megabreccia (36.8765, -117.4998). A massive landslide shed from the faulted western side of Tin Mountain provided the breccia.

Racetrack Valley Road is at an elevation of nearly 5,000 feet here. After passing through a forest of abundant Joshua trees—a spiny yucca of the agave family—on the fans to the east, the road begins a long descent to Teakettle Junction, which is about 20 miles from the pavement at Ubehebe Crater. In another mile or so the Racetrack comes into view.

The playa's name comes from its roughly oval shape and the unusual bedrock islands near its north end: The Grandstand (73 feet high) and a smaller neighboring carbonate-rock knob. By a stretch of the imagination, these islands could be a grandstand for a cheering crowd and a judges' stand for Roman chariot races. The playa occupies the south end of Racetrack Valley, between two prongs of the Last Chance Range and Cottonwood Mountains, at an elevation of about 3,708 feet. It measures 2.8 miles north to south and up to 1.3 miles east to west, and its northern end is 3 feet higher than its southern.

At its southern end the Racetrack lies directly against a steep, 850-foot-high mountain face of dolostone. Cobble- to boulder-size stones loosened by processes including frost heaving, animal action, plant growth, earthquakes, and precipitation tumble down the cliff face and are strewn on the playa surface. Most of the stones are irregular fragments of dark dolostone.

Mud cracks cover most of the Racetrack's surface. They are old, semipermanent features outlining polygons 3 to 4 inches in diameter that are bounded by cracks about an inch deep. They form as the playa's mud surface dries and shrinks, radiating from regularly spaced centers, usually in sets of three cracks at 120 degrees to each other. Although not as regular as polygons at Devils Postpile (Vignette 31), those at the Racetrack are far more regular than desiccation cracks that form in mud after it rains. Staring at the mud cracks as you walk across the playa can be mesmerizing. There sure are a lot of them! If we idealize the polygons as hexagons with side lengths of 2 inches, then the area of each is a little more than 10 square inches. The area of the playa is about 3.7 square miles, so that's about 1.5 billion hexagons.

The southern end of the playa is littered with thousands of angular dolostone blocks derived from the hill behind the photographer. View is northwest.

Some of the billion or so polygonal mud crack polygons on the playa surface. Polygons are typically 3 to 4 inches across.

Stones have slid, somehow, across the playa surface, leaving distinct trails in the wet mud tens to hundreds of feet long, a few inches to 12 or more inches wide, and a small fraction of an inch deep. Unless abnormally deep, tracks are discernible for no more than three or four years.

Walk out onto the playa's surface so you can examine the various trails. You will likely find a pile of mud in front of the stones, bulldozed ahead during movement. Low levees—rounded ridges of dried mud a small fraction of an inch high—border some of the trails. You may then conclude that as a stone plowed across the muddy playa surface, it pushed mud aside and ahead of it.

On the south part of the playa, the prevailing direction of stone movement is to the north-northeast, but the trails commonly diverge 30 degrees or more to either side of north, and some of the trails reverse direction. Long trails form in segments, indicating that the stones moved episodically. Other curiosities that you will encounter on the playa are well-developed tracks that look as if they were made by sliding rocks, complete with little bulldozed piles of mud but lacking rocks at their termini. Although some rock-less trails may lack a terminal rock because someone moved it, we invite you to ponder other ways in which such a trail could be made.

The next chapter in the saga of the slithering stones began in 2010 when cousins Richard and Jim Norris (a geologist at Scripps Institution

Stones on Racetrack playa are mostly angular blocks of dolostone, which have slid across the playa when it was muddy, bulldozing some mud in front and making short levees on either side of the trail. —Photo by Marli Miller, geologypics.com

VIGNETTE 14: THE RACETRACK   117

A mystery: clear trails with levees and bulldozed piles of mud, but no rocks. —Photo by Jim Norris

of Oceanography and an engineer at Interwoof, respectively) assembled a team to track moving stones. Under permits from the National Park Service, they installed special GPS trackers in fifteen rocks gathered outside the park, then placed these tracker rocks on a part of the playa surface without abundant natural stones. Nails driven in the playa beneath the rocks kept magnetic switches in the off position. If moved off the nails, the circuit would close, and the GPS units would begin tracking. They also set up a weather station and a time-lapse weather camera. They were ready to track rocks during the infrequent, intense storms that were suspected of moving them—storms so severe that no observers would be around to see the rocks move.

In November 2013, 1.4 inches of rain and about 8 inches of snow fell, and a shallow pond formed on the playa. Owing to the southward tilt of the playa surface the pond was confined to the southern end. The temperature dropped at least a few degrees below freezing nearly every night. The team arrived on December 18; it snowed on December 19, and the pond had a thin sheet of ice on it.

The next day, sunny and clear with light winds of about 10 miles per hour, the stones moved. Sunshine caused floating ice to break into floating panels, some initially 50 feet or more across, and as ice melted and open water formed between the panels, the wind moved them, and they, in turn, pushed stones ahead of them. The research team was on the north end of the pond in the morning, but by the afternoon of December 20

Aerial view of standing water on Racetrack playa on December 26, 2013. The pond in the southern part of the playa (closest to the viewer) was no more than 4 inches deep (lighter colored water), shallowing to zero through the dark-brown zone. The north end of the playa was damp but did not have standing water. —Photo by Mike Hartmann

Parallel tracks in the mud formed during the December 20, 2013, event. Only light winds were required to move thin floating ice sheets that in turn pushed the rocks. By the afternoon, when this photo was taken, the ice had melted and most of the water had been pushed north, away from the southern shore, exposing the new tracks. —Photo by Jim Norris

steady light winds had pushed most of the water off the south end of the playa and they beheld newly formed, parallel trails in the mud.

Three more episodes of movement happened in late December and early January, and on January 9 the team witnessed and filmed a stone surrounded by an isolated ice sheet move past its neighbors. Caught in the act! As before, the day was sunny with light to moderate winds, and movement occurred after the thin ice covering the pond had largely broken up. The mystery had been solved: stones move on relatively pleasant afternoons, when morning sun has partially melted the thin ice sheets covering the pond and light winds push the ice around.

Some of the rocks that had been equipped with GPS receivers did their jobs, recording track directions and speeds. Others were disabled by moisture that leaked into the electronics compartments. Rocks moved slowly, generally no more than 20 feet per minute. Some of the instrumented rocks moved up to 735 feet between December 2013 and January 2014, in multiple events.

A big surprise was the thinness of the ice involved in the rock movement. One hypothesis for how ice could move rocks relied on thick ice sheets to encase the rocks and float them partially off the muddy bottom, but the ice that moved the rocks was just 2 to 5 millimeters thick. Ice breakup was clearly important because a full ice sheet covering on the pond, anchored at the shore, would not move. Ice melted around the rocks first because they absorbed the sun's rays more efficiently, and once large panels were freed from their neighbors, they were able to float freely and push rocks.

The observations of thin ice pushing rocks around provided answers to the conundrums posed above. The ice commonly fragmented upon contact with rocks, and ice panels sometimes stacked up on the upwind sides. This could explain how the rock moved out of Sharp and Carey's corral. The rock that did not move out of the corral had a stake on the immediate upwind side that could have broken up any ice slabs that were on their way into the corral to herd stones out of it.

You may be surprised to learn that not everyone is pleased that the Racetrack mystery has been solved. People love mysteries, and what fun is a National Park, particularly Death Valley National Park, without geologic mysteries to contemplate and solve? The Norwegian geochemist Tom F. W. Barth was quoted as saying, "Let us leave some mysteries for our children to solve, otherwise they shall be so bored." We are pleased to report that there are countless perplexing geologic mysteries yet in Death Valley and eastern California for future generations of geoscientists to study. This book is full of them. For starters, you might try to come up with an explanation for the rockless trails.

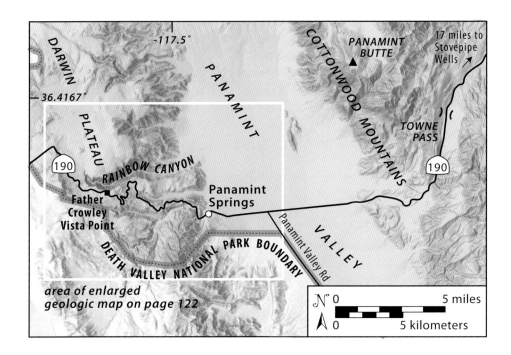

GETTING THERE

Father Crowley Vista Point is alongside CA 190 near the western entrance to Death Valley National Park. The overlook lies 38.5 miles west of Stovepipe Wells and 7.8 miles west of Panamint Springs. Travelers from the west will find it 42 miles east of Lone Pine. The parking area provides a splendid view of Rainbow Canyon, but the Panamint Valley overlook, 0.6 mile to the east on a drivable but bumpy dirt road, looks into a more intriguing part of the canyon and gives an unobstructed view of northern Panamint Valley and Panamint Butte, east of the valley. Because the climb from Panamint Springs gives close highway views of the various geologic units that you can see at the stop, the discussion below includes a brief road log, starting in Panamint Springs. Beware of high-speed traffic on CA 190 and park only in wide pullouts. Vault toilets are available at the parking area.

VIGNETTE 15

RAINBOW CANYON AND FATHER CROWLEY VISTA POINT
A Sermon on the Basin and Range

The transformation of Death Valley National Monument into Death Valley National Park in 1994 was accompanied by significant expansion of protected lands on its west side. Much of the area between Saline Valley and the Argus Range is rarely visited, yet each year hundreds of thousands of people drive the stretch of CA 190 that goes through it. Father Crowley Vista Point, an overlook on this route atop the steep western drive out of Panamint Valley, gives a dramatic view of the complex geologic history of this area, and Rainbow Canyon, a 1,500-foot-deep gorge adjacent to the parking area, is both geologically stunning and a venue for low-altitude military jet aircraft training. You may find your geologic reveries disturbed by an F-18 roaring out of the canyon on full afterburner. There is plenty to see at the parking area, but a drive or stroll to the Panamint Valley overlook itself, 0.6 mile down the bumpy dirt road to the east, is highly recommended.

Our tour begins in the hamlet of Panamint Springs at the mouth of Darwin Wash on the west side of Panamint Valley. Set your odometer to zero at the motel, and we will introduce the principal players as we drive to the overlook. The road swings southwest for half a mile before hooking right around a spur of pinkish-brown granodiorite of Jurassic age, about 170 million years old. After the hook, you are pointed north at dark basalt lava flows, about 4 million years old, that overlie almost everything in this area. Here they sit on granodiorite at an elevation of about 2,200 feet; tan material that looks like limestone, obvious where the road swings westward, is tufa from pluvial Lake Panamint. The road climbs past roadcuts in tan fanglomerate, locally covered by copious basalt cobbles. Watch for a large turnout on the right at mile 1.7. Park and carefully walk back down the road to see three basalt dikes cutting the fanglomerate. These are clearly dikes rather than tilted lava flows because they cut layering in the fanglomerate and have orange zones on each side where the hot magma baked the host rocks. We will see many more dikes across the canyon at the overlook.

Continue uphill on CA 190. The road soon passes into roadcuts in large basalt blocks cemented by white caliche (calcite precipitated by

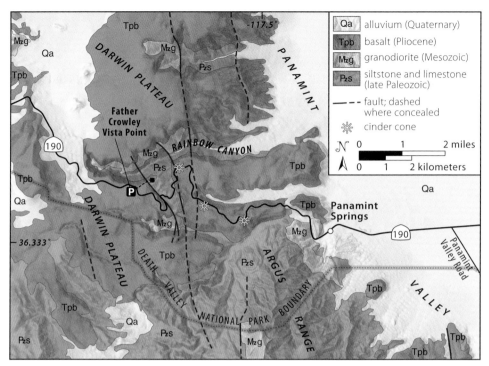

Simplified geologic map of the eastern Darwin Plateau. The road passes through three cinder cones and crosses several faults on its way to the overlook. —Modified from Hall and MacKevett, 1962

Basalt dikes cutting fanglomerate along CA 190 at mile 1.7. The reddish color of the rocks adjacent to the dikes was caused by the oxidation of iron by hot gases emanating from the dikes when they intruded. The dike on the right is about 3 feet thick.

evaporating groundwater), and into bedded reddish cinders—basaltic tephra—and thin basalt flows. Here the road was cut through a cinder cone, one of the vents for the basalt lavas. The road wraps left around the cinder cone with its red cinders and fragmental gray lava in the roadcuts. Beautiful beds of red and orange basaltic tephra are exposed in a double-walled roadcut at mile 2.9. The road and scenery flatten over the next few miles. Watch for a large pullout on the right at mile 5.2 where the road enters another cinder cone. Stop for a close look at the interior of a cinder cone and the variety of basaltic tephra.

Typical basaltic tephra in roadcut at mile 5.2.

At mile 5.9 the bedrock character changes dramatically, from fanglomerate, basalt lava, and basalt cinders to light-tan metamorphosed limestone and siltstone. We have crossed a fault and temporarily passed out of the basalt field into marine deposits of Pennsylvanian or Permian age (about 300 million years old). These are well exposed after mile 6.3, but available parking is limited. At mile 6.6, shortly after a sharp left hairpin turn, you pass a dike, one of several cutting siltstone beds along this stretch. This deeply weathered dike of the Late Jurassic (150-million-year-old) Independence dike swarm contrasts sharply with the much younger Pliocene dikes seen at mile 1.7.

At mile 6.7 the road reenters basalt until the turnoff into a large parking area for the Father Crowley Vista Point at mile 7.9. The paved parking area is at the edge of Rainbow Canyon, here about 1,100 feet deep. The north wall exposes a panoramic view of the three geologic units that

A deeply weathered Jurassic dike, a member of the widespread Independence dike swarm, cuts light-colored siltstone.

we drove through on the climb from Panamint Valley. The lower half of the ridge on the north side of the gorge consists of steeply dipping layers of tan limestone. Behind and above them is granodiorite like that encountered west of Panamint Springs, knobbly and dark in shades of brown and green. The basalt flows, here totaling about 400 feet thick, flowed across a flat surface on the granodiorite and rim the canyon. If the nearby cinder cones erupted today, the basalt would cascade into the deep canyon and flow into Panamint Valley.

These three rock units represent the three main geologic events in the history of western North America. The limestone was deposited in Paleozoic time during the last stages of marine deposition off the western continental margin, in a setting similar to that of the east coast of North America today (but with far different flora and fauna!). This time of quiet sedimentation, perhaps 300 million years long, terminated around 230 million years ago in the Triassic Period, when complicated plate rearrangements turned the passive margin into a subduction zone. The granodiorite was intruded as part of a magmatic arc, akin to the modern Andes, produced by subduction. The Sierra Nevada batholith, large granite bodies in the Inyo Mountains, and the Hunter Mountain batholith at the north end of Panamint Valley are part of this event. The

VIGNETTE 15: RAINBOW CANYON AND FATHER CROWLEY VISTA POINT   125

View north from the parking area. Layered basalt rims the northern edge of Rainbow Canyon. The layer-cake nature of the lava flows suggests that down-dropping of the valley and erosion of the canyon happened after the basalt erupted. A thin orange zone, probably a soil horizon baked and oxidized by the lava, separates the basalt from the underlying granodiorite. The contact between the granodiorite and limestone (tan rock at lower right) is steep and oriented approximately east-west.

basalt is representative of widespread volcanism that accompanied Miocene and younger crustal extension.

The intrusive contact between the limestone and granodiorite is irregular and wanders up and down the canyon wall, but the lower contact of the basalt on granodiorite is relatively planar, showing that the lavas erupted across a low-relief landscape eroded across the granodiorite. This basalt-on-granodiorite contact, marked here by a thin orange baked zone, is at an elevation of about 3,900 feet, far above the basalt-granodiorite contact near Panamint Springs (elevation about 2,200 feet). The paucity of baked zones within the stack of basalt flows suggests that they erupted within intervals of time too short for soil to form.

Drive eastward down the dirt road, or walk along a trail at the canyon edge, toward the overlook. You will see that the lower contact of the basalt steps down along faults, part of the stretching process that opened Panamint Valley. The drive up from Panamint Valley crosses several of these faults.

The overlook, at 4,175 feet, has an unobstructed view across the north end of Panamint Valley with its playa on the valley floor at 1,500 feet and a small field of sand dunes at its north end. Panamint Butte is a huge, dramatically striped mountain capped by dark basalt that

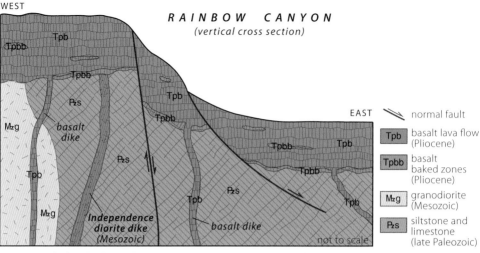

Idealized relationships among rock units, dikes, and faults in Rainbow Canyon.

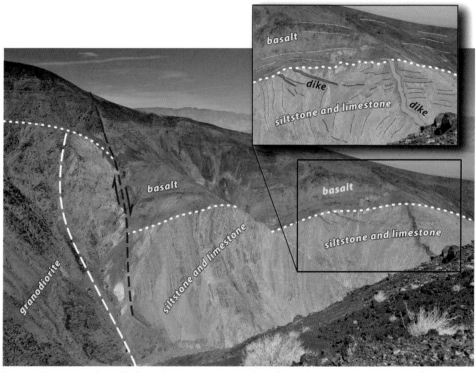

The complicated north wall of Rainbow Canyon, seen here from about 1 mile northeast of the overlook and 1.7 miles east of the parking area, exposes gray granodiorite on the left intruding tan limestone and siltstone, here inclined vertically, along a steep contact (dashed white). Both units were planed off along a relatively flat erosion surface (dotted white) before lava flows covered the land. A steep fault (dashed red) that trends north dropped the basalt contact 600 feet down to the east. This normal fault is one of a series that dropped Panamint Valley down along the east side of Darwin Plateau. The Paleozoic sedimentary rocks were intruded by numerous basaltic dikes of both Mesozoic and Pliocene age. Inset shows interesting dike relationships on the right side of the photo. There are two generations of dikes. Dikes marked in blue stop at the eroded contact that the basalt flows sit on and may be Jurassic in age; Pliocene dikes (green) cross the contact and cut the overlying basalt flows, some of which they fed. Other thin dikes are evident in red cinder areas elsewhere in the north wall.

View eastward across Panamint Valley to Panamint Butte (skyline, center) and gray ridgeline of granodiorite on the left. The layered gray Paleozoic sedimentary rocks beneath Panamint Butte are truncated by granodiorite along a steep intrusive contact (dashed). That relationship, together with the overlying dark basalt flow, is the same as that exposed in Rainbow Canyon (bottom left) but they are displaced from one another along the Hunter Mountain fault. The red rocks in the center foreground are tephra from one of the cinder cones cut by the highway.

resembles a giant multilayer cake with chocolate frosting, although the cake was folded like a taco, then had its lower layers cut off and placed on top, and then was diced by a sharp knife before the frosting was applied. Granodiorite of the Hunter Mountain area on the left is in steep contact with thinly bedded Paleozoic sedimentary rocks on the right, all capped with those dark 4-million-year-old basalt lava flows, the same geologic relationships we see here in Rainbow Canyon.

The prominent basalt cap to Panamint Butte is indistinguishable in age and composition from that beneath your feet and on the other side of Rainbow Canyon. All are 4-million-year-old basalts with visible crystals consisting of rusty, 2- to 3-millimeter-long olivine, sparse black clinopyroxene up to 4 millimeters long, and sparse plagioclase. These lavas, erupted mostly from the cinder cones you drove through on the way up here, are not present in Panamint Valley, and their layer-cake aspect implies that they erupted and flowed across a flat surface, not into canyons that dropped into a valley. The most reasonable explanation for this arrangement is that the lavas erupted across a plateau that was continuous with the Panamint Range prior to existence of Panamint Valley. The implication of this explanation is that Panamint Valley opened and has widened at least 5 miles in the last 4 million years. That's only an average rate of 0.1 inch per year and is consistent with GPS measurements.

Aerial view of Panamint Butte. The black and white layered rocks are mostly thinly bedded late Paleozoic sedimentary rocks like those at Rainbow Canyon, tightly folded on the left and diced by faults. Thicker gray and white beds in upper left, under the dark basalt cap, are much older early Paleozoic Bonanza King dolostone, placed here by a thrust fault (red dashed line) that is responsible for their folding.

FATHER CROWLEY

Father John J. Crowley, a legendary Catholic priest, was one of the most influential advocates of tourism in an area desperately in need of economic development during the Great Depression. He was born December 8, 1891, in County Kerry, Ireland. After completing seminary studies in 1919, the "Desert Padre" served a desert parish for five years covering 30,000 square miles from Bishop to Barstow. After spending ten years at a parish in Fresno, California, Father Crowley returned to the Owens Valley in 1934 to find that the diversion of the Owens River to Los Angeles had turned a once verdant valley of flourishing farmland into desert. With the demise of agriculture as an economic base, Father Crowley believed tourism would save Owens Valley and its poverty-stricken residents. He devoted the last six years of his life publicizing the hunting, fishing, skiing, and some of the most spectacular scenic views in the nation. In one publicity effort, Father Crowley convinced Inyo County to declare opening day of trout season a county holiday, and he blessed fishing equipment every year. In another, he climbed Mt. Whitney on Sept. 14, 1934, and became the first priest to celebrate Mass on the summit. He died tragically in a single-vehicle automobile accident in 1940.

When the new Long Valley Dam was completed in 1941, the reservoir it created was named Lake Crowley in honor of the visionary desert priest and his unceasing efforts to publicize the eastern Sierra as an ideal tourist attraction.

Although basalt lava flows blanket most of this side of the valley, geologists have traced the contact of granodiorite and Paleozoic sedimentary rocks from Rainbow Canyon northward for several miles to where it bends eastward to emerge into the north end of Panamint Valley near the sand dunes. There, the contact is separated 5 miles in a right-lateral sense from its position beneath Panamint Butte. That separation was achieved by displacement along the Hunter Mountain fault.

Notice on the map how the Hunter Mountain fault is bent like a lazy "S" and that as right-slip accumulates on the central section of the fault, the Argus Range pulls away in a northwest direction from the Panamint Range to open Panamint Valley. But this fault does double duty, for at the same time, the Inyo Mountains pull away from the Cottonwood Mountains, and Saline Valley also opened as a valley almost as long and wide as Panamint Valley, and equally as deep. Thus, Saline and Panamint Valleys are extensional basins linked by the Hunter Mountain fault and formed by crustal stretching.

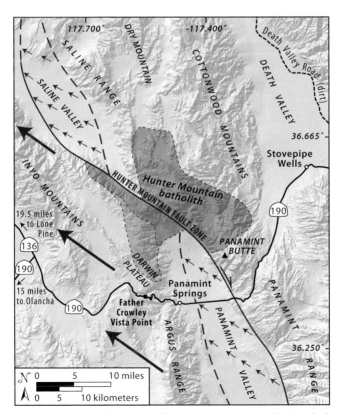

The southeast end of granites of Hunter Mountain is displaced about 5 miles right laterally along the Hunter Mountain fault zone. The area within short dashes is the inferred extent of the Hunter Mountain granite beneath lava flows on the Darwin Plateau.

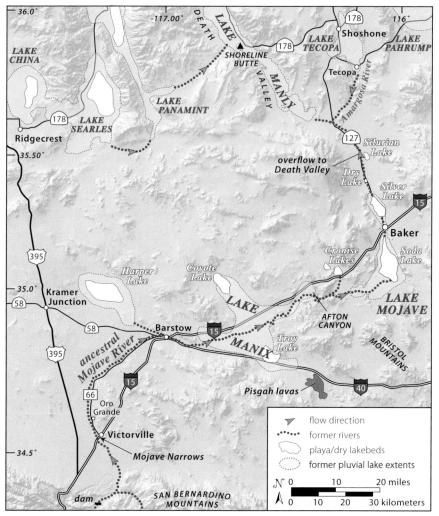

The modern course of the Mojave River and related features discussed in this vignette.
—Modified from Enzel and others, 2003

GETTING THERE

Travelers to Death Valley on I-15 proceed to Baker and then go north on CA 127. The route follows a course closely parallel to the channel of the glacial-age Mojave River, when it flowed from snow-covered 8,000-foot peaks in the San Bernardino Mountains to Death Valley, the ultimate sump of California deserts at 282 feet below sea level. Features along the river's former route can be viewed along I-15 and CA 127 on the way to Death Valley, some from main highways and others by way of short detours.

To view the narrows at Victorville, travelers northbound on I-15 can exit for CA 18 east at exit 153A toward Apple Valley. Make a right turn onto D Street and drive 0.6 mile to the train station at 6th Street. Turn left, park, and look south down the

VIGNETTE 16

THE MOJAVE RIVER
A Waterway in Search of a Drain

The San Bernardino Mountains, southern California's highest range, spawn two good-size rivers, the Santa Ana and the Mojave. The Santa Ana is well behaved and flows 65 miles southwest from the mountains into the Pacific Ocean near Newport Beach. The other, the Mojave, is an oddball: it flows almost directly away from the ocean, and during times of heavy precipitation it flows north and northeast 135 miles from the San Bernardino Mountains to Silver Lake playa, about 5 miles north of Baker. In glacial times, roughly the last million years, ice and deep snows in the San Bernardino Mountains fed the Mojave River, which at times flowed all the way to Death Valley, a total journey of around 200 miles. Along the way, the river's water cut deep rock gorges, languished for months in placid lakes, overflowed into other basins, and provided food and attractive lakeshore campsites for Native Americans as well as habitats for fish, birds, and mammals.

Millions of years ago streams in the Mojave region did what they were supposed to do: they carried sediment southwest toward the sea. But the modern San Bernardino Mountains started to rise about 2.5 million years ago, causing some of these streams to flow north into a basin in the Victorville region. About 750,000 to 500,000 years ago, this latter system of streams and lakes overflowed eastward and carved a shallow canyon at what is now Barstow. The river then cut and spilled through another granite ridge east of the Barstow area, eventually continuing

railroad tracks to the narrows. Travelers wishing a closer view of the narrows can return to D Street, turn left onto D Street/CA 18, and go another 0.5 mile to the river crossing. This high-speed road provides no place to stop, but it gives a close view of the narrows.

Reverse course and drive north on CA 18, cross the Mojave River on a steel truss bridge, and pass the Oro Grande cement plant, where marble is processed to make cement. This pleasant back way to Barstow follows the river and the railroad. Look for bluffs cut by the river across from a solar farm about 8 miles north of the Mojave River bridge.

to Death Valley. It is no mean feat for a river of modest size and limited water to forge a 165-mile-long channel across a desert. The Mojave River took advantage of its origin in lofty mountains along the San Andreas fault and simply obeyed gravity, traveling in search of lower topography, going underground when necessary, and ending up at the lowest dry land on the North American continent.

Let us follow the course of this unusual river, beginning where it emerges from the Mojave River Forks Dam, a flood-control structure at the northern base of the San Bernardino Mountains. The riverbed beyond the reservoir is exceptionally wide and sandy, so water flowing along it, unless in flood quantity, generally percolates into the sand. Underground flow is an efficient mode of water transport, especially in deserts where evaporation is high and surface runoff is often damaging. Most downstream water users drill wells to tap into the slow-moving subsurface groundwater, which thus serves as a storage system far less expensive and more efficient than any artificial reservoir. All this is hard on fish, though, and beyond the foot of the mountains, the Mojave River is not favorable fish habitat.

During summer when most of the riverbed between the mountains and Victorville, 15 miles downstream of the mountains, appears bone dry, surface water flows continuously through the bedrock narrows in the center of Victorville. There, the river flows within a steep-walled gorge 140 feet wide, 150 feet deep, and 1,000 feet long cut into a rocky granitic spur projecting west from the mountains.

Railroads use the gap that the Mojave River has conveniently cut for them across the formerly buried granite spur at Victorville (looking north).

Why does water flow on the surface in the Mojave Narrows but not upstream from it? Along the upper Mojave River, water-saturated sands lie close to the riverbed surface. The subsurface water flows slowly through the sands until it meets impervious granitic bedrock, which forces the water to the surface. The thin veneer of channel sands in the narrows cannot accommodate all the water so it flows on the surface year-round for a mile or so downstream as far as the I-15 bridge. The water percolates back into the sandy bed north of the I-15 bridge but returns to the surface 3.5 miles above the Oro Grande cement plant, situated on Old Route 66. There, the river flows through a second bedrock narrows cut through a formerly buried spur of granitic rocks that projects west from Quartzite Mountain. The water usually disappears again within a mile or two, but it flows close to the surface in places, such as Palisades Lake and Silver Lake Roads, 4.5 miles downstream from the Oro Grande cement plant. The groundwater table is shallow all along the riverbed, as little as 20 feet below the surface in places, but it is deeper farther from the main channel. Subsurface Mojave River water, pumped mainly from wells, sustains a broad belt of farmland and habitation from Victorville to beyond Barstow.

Between Victorville and Barstow, 35 miles by river, the streambed is mostly wide, sandy, and dry. Old Route 66 closely parallels the river and carries less traffic than I-15.

At Barstow the river and I-15 turn northeastward. In glacial times, the river encountered its first ponded water in Lake Manix, about 25

Looking downstream (northeast) at the 250-foot-wide, sandy bed of the Mojave River at the Hinkley Road bridge, about 14 miles upstream from Barstow. Water is not far below the surface.

miles northeast of Barstow. The ancient lake covered about 85 square miles to a maximum depth of nearly 200 feet. For a fuller account of Lake Manix, see *Geology Underfoot in Southern California*.

Lake Manix's initial outlet probably drained east-southeastward down the wide trough extending from Barstow to Bristol Lake playa near Amboy and then on to the Colorado River near the present-day town of Blythe. Currently, the Burlington Northern Santa Fe Railroad, I-40, and Old Route 66 closely parallel the river's ancient, well-graded route as far as Amboy.

Rivers are adept at seeking out low areas within rough terrain and have a clever way of finding the lowest path around an obstacle. They simply form a lake, and the water level within this impoundment rises until it reaches the lowest outlet. The weakness of this procedure is that the lowest point does not necessarily lead to the lowest areas within the terrain beyond. In the case of the Mojave River, the lowest point nearby was Death Valley, not the Colorado River and the ocean that it empties into.

According to Brett Cox and his US Geological Survey colleagues, the Mojave River's course changed dramatically about 25,000 years ago. The southeastern course was blocked either by lava flows from Pisgah

The Mojave River encounters impermeable bedrock and flows on the surface in scenic Afton Canyon, requiring the railroad to cross the river on a bridge where the river swings sharply to the right to go around a horseshoe bend before resuming its eastward course to lower ground of the Devils Playground, south of Baker. View is downstream and east toward the Bristol Mountains on the skyline.

Crater or by displacement on the Pisgah fault. The blockage caused more water to feed into Lake Manix, which rose until it overtopped its northeast rim. The large outflow soon cut a narrow, 5-mile-long gorge at what is now Afton Canyon through which the Mojave River extended its course toward Death Valley.

Railroads prefer water-level routes, and the Union Pacific Railroad currently follows Afton Canyon on its way to Las Vegas. Water flows year-round through parts of Afton Canyon for the same reason that it does so at Mojave Narrows: impermeable bedrock in the channel floor.

One easy-to-see remnant of the ancient lake is a magnificent beach ridge at Afton. From I-15, take the Afton off-ramp (exit 221 about 40 miles east of Barstow) and park at the top of the ridge about 200 yards south of I-15. The level, rounded crest of the ridge and the many flat, smooth beach pebbles on its flanks reflect its origin. Strong winds from the west created huge waves on Lake Manix, and the breakers picked up and reworked sand and stones along the shoreline. See Vignette 8 for photos of a similar beach ridge that formed along a Lake Manly shoreline in Death Valley.

The river's profile reflects its recent history. Upstream of Afton Canyon the river drops at a steady, gentle gradient of about 16 feet per mile. There must have been a sharper drop around Barstow when the river system first breached that barrier to flow east to Afton Canyon, but any trace of that kink in the river's profile has been eroded away in the subsequent half-million years. The cutting of Afton Canyon was much more recent, only about 24,500 years ago, and the gradient in that canyon is much steeper. It flattens in the broad expanse of the Devils Playground, south of Baker, then drops steeply from Silver Lake to Silurian Lake north of Baker and on to the Amargosa River and Death Valley.

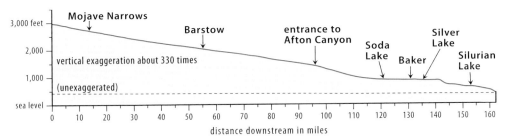

Downstream profile of the Mojave River, from the Mojave River Forks Dam at the foot of the San Bernardino Mountains to its junction with the Amargosa River north of Baker. The river is well graded upstream of Afton Canyon and then drops in a series of steps through Afton Canyon and over the Silver Lake outlet to its junction with the Amargosa River. Note the extreme vertical exaggeration; the dotted line gives the slope of the unexaggerated profile—pretty flat.

A wave-cut cliff near the base of the hill west of the north end of Silver Lake playa. Wind-blown silt emphasizes the strandline and fills in gullies. The prominent horizontal notch near the base of the hill, with largely sand-covered outcrop below it, corresponds to the level of Silver Lake's outlet and was carved when Lake Mojave overflowed to the north in the cooler, wetter climate of the Pleistocene Epoch, reaching Death Valley.

Upon emerging from Afton Canyon, the Mojave River flows east on a broad, gently sloping alluvial plain, much of which the river built by depositing its sediment load. There, its channels are shallow, braided, and constantly shifting. Some floodwater occasionally flows north into Cronise Basin, passing under I-15 at the Basin off-ramp (exit 230).

The largest floods on the Mojave River send water farther east into Soda Lake playa, which connects northward to Silver Lake playa by a channel through the center of downtown Baker. At least once a decade the water goes all the way to Silver Lake, inflicting considerable discomfort and flood damage to Baker. The greatest historical water depth recorded at Silver Lake was 10 feet in 1916, an unusually wet year, but several feet of water have accumulated in other years. Water depth in Silver Lake must be above 35 feet for it to overflow, which it did in glacial times. Baker would have been largely submerged then, except for the upper part of its giant thermometer and a few signs, and Silver and Soda Lakes were joined to form a larger body, Lake Mojave.

Native Americans found the shores of Lake Mojave an attractive place to live, and evidence of their occupation has been found in numerous places along the now-abandoned shorelines. From CA 127 along Silver Lake's east side, you can see traces of old shorelines cut into the base of hills west of the lake, especially when shadowed in the late afternoon.

After Lake Mojave overtopped the basin-confining bedrock ridge at Silver Lake, the water continued north toward Death Valley, ponding at what is now Silurian Lake playa (*Silurian* is its name, not its geologic age). From Silurian Lake, water flowed north into a minor pond nestled against the eastern base of the Salt Spring Hills. Here, once again, water overflowed a saddle in a narrow ridge of granitic rock and cut a narrow gorge. A cluster of salt cedar trees, which benefits from ponding of groundwater behind the granitic spur, 0.5 mile east of CA 127, is accessed from a short interpretive trail. Once across this last spur, the Mojave River had an easy run of about 3 miles across alluvium to a junction with the Amargosa River, which still flows into Death Valley. Waters of the combined rivers contributed greatly to Lake Manly in glacial times (Vignette 8).

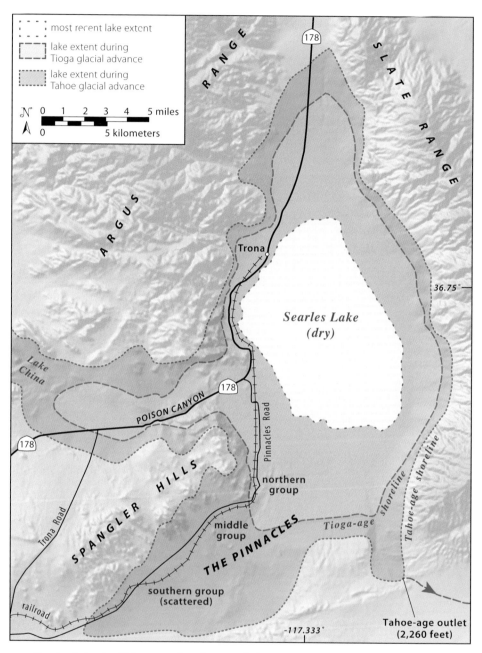

Searles Lake, its Pleistocene shorelines, and the pinnacle groups. During its highest water level, the lake overflowed to the southeast and into Panamint Valley.

VIGNETTE 17

TRONA PINNACLES OF SEARLES LAKE
An Alien Landscape

The Trona Pinnacles, one of the most unusual landscapes in all North America, resembles a cartoonist's rendition of the Moon's surface before the Apollo astronauts showed otherwise. The pinnacles have appeared in numerous movies, television shows, and commercials, commonly to indicate an alien landscape. They are reached via a 7-mile detour on a well-graded gravel road that branches off CA 178, a lesser-used but quite rewarding route through Searles and Panamint Valleys into central Death Valley.

Lake Searles was the third lake, after Owens and China, in a string of five major pluvial water bodies nourished mostly by the Owens River, which carried water from the melting ice and snow of the Sierra Nevada to Death Valley in Pleistocene time (see map on page 62). At its maximum, when it was 800 feet deep, Lake Searles overflowed its southern margin into the south end of Panamint Valley and then north into Lake Panamint, which was even deeper (1,000 feet) at its maximum. Shoreline features such as wave-cut cliffs and benches, gravel bars, beaches, tufa pinnacles (about which shortly), and lakebed deposits identify two principal phases in the late history of Lake Searles. These remnants of the ancient lake have been correlated with the last two major glaciations in the Sierra Nevada, the Tioga (about 20,000 years ago) and Tahoe (about 140,000 years ago).

GETTING THERE

CA 178 traverses the western shore of Searles Lake playa in the last part of a 25-mile drive from Ridgecrest to Trona. The dirt road to the pinnacles turns southeast off CA 178 about 16.6 miles west of the center of Ridgecrest, 0.5 mile before the road swings northward around a bedrock ridge split by a dike that bears a resemblance to the spiked back of a stegosaurus. Signs mark the junction, and initially the road is wide and well graded. Follow it 1.3 miles to a railroad crossing and continue south another 3 miles to a flat-topped, linear gravel ridge, yet another beach ridge like those found at pluvial Lake Manix and Lake Manly. Continue over the ridge to the pinnacles area. After wet weather, soft, slippery lake clays can make the road impassable beyond the railroad crossing. Only the northern group of pinnacles is comfortably accessible by passenger cars.

VIGNETTE 17: TRONA PINNACLES OF SEARLES LAKE

Aerial view of pinnacles in the northern group, looking southeast into the Mojave Desert.

The five large interconnected pluvial lakes functioned as a succession of huge decanting vessels that let much of the mud from the Owens River settle out, passing clearer water downstream. The water became progressively enriched in soluble salt compounds down the chain, as evaporation from preceding basins concentrated the elements. Lake Searles lay in just the right place to accumulate unusually rich and varied deposits of salts, including substances of commercial use. Searles hosts a large operation reclaiming and processing valuable chemical compounds, many containing the element boron in the form of borate minerals. Ownership of the resources and facilities at Searles Lake, mainly in Trona, has passed through several hands and as of 2021 rests with Searles Valley Minerals Inc., a subsidiary of the Indian conglomerate Nirma.

The company obtains salts mostly by drilling shallow wells and pumping brines to the chemical plants rather than by scraping the lake floor. Cores from a huge number of drill holes provide valuable subsurface geologic information, as well as organic substances that can be dated by radiocarbon. Whenever the lake dried up, evaporation concentrated a layer of salt on the lake bottom. When incoming water poured into the basin and made a new lake, it laid down a layer of mud. The resulting deposits are interbedded layers of salt and mud.

Approximately five hundred pinnacles are clustered into three groups. The tallest pinnacle, about 140 feet tall, is in the middle group. The

BORON

Modern technology depends on several elements that are extremely rare on Earth. Lithium, for example, is critical to modern high-energy, lightweight batteries; LEDs depend on several scarce elements, including yttrium and gallium; and batteries for electric cars use large amounts of rare earth elements such as neodymium. We rely on geologic processes to concentrate these rare elements into deposits that are rich enough that they can be extracted economically. Boron, which is used in a wide variety of products including glass, cleaning products, and drugs, is an important product at Searles Lake and has been in Death Valley.

Boron is the fifth-lightest element and is extremely rare in the universe because element-making processes in stars, which turn hydrogen and helium into heavier elements, skip a few—lithium, beryllium, and boron—and jump straight to carbon. Instead, boron forms when high-energy cosmic rays hit the nucleus of an atom, such as a nitrogen atom in the atmosphere, and knock off a small piece. Much of this ends up in dust, which ends up in the ocean, and nearly all oceanic crust is eventually subducted. Magmatic activity in subduction zones scavenges boron. Hot springs such as Hot Creek (Vignette 29) in Long Valley caldera tend to be rich in it, and some of the boron in Searles Lake sediments is thought to have come via the Owens River from hot springs there.

Dissected lakebed deposits of the combined Lakes Searles and China are exposed in the headwaters of Poison Canyon along the north side of CA 178, east of the road to Red Mountain.

pinnacles formed in relatively shallow water along the west shore of a large bay on the southwest side of Lake Searles. They are crowded close to the east and southeast flanks of the Spangler Hills, a mass of hard, jointed granitic rock that extends under the dry lake and is covered by a relatively thin mantle of lake-bottom deposits. A succession of strandlines on the east face of the Spangler Hills, at elevations well above the tops of the northern group of pinnacles, shows that the pinnacles must have been deeply submerged by rising lake levels, if they existed when the strandlines were made. You can best see these former shorelines from the north pinnacles area when a low western sun backlights them and creates shadows on the wave-cut cliffs.

Pinnacles are made of tufa, which is principally calcium carbonate. Most of the tufa towers are circular in cross section, reasonably symmetrical, and usually 30 to 40 feet high; some are more than three times as tall. Tubby structures 20 to 30 feet high, with an elongated elliptical shape, are called tombstones. Small, dumpy tufa cones, mostly 10 to 15 feet high, and long linear tufa ridges with serrated crests are common. The northern group's most prominent tufa ridge is 500 feet long, highly serrate, and includes a 120-foot-tall pinnacle. Most pinnacles rise from a broad, gently sloping base of fallen tufa blocks mixed with beds of lake sediment containing lenses and layers of tufa.

The pinnacles did not all form at the same time. Tioga and Tahoe glacial waters submerged the middle and northern groups, but pinnacles of the southern group stand at higher elevations and could have formed only in a pre-Tahoe-age lake. Much deteriorated by weathering and

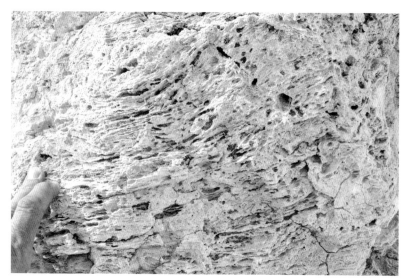

Tufa, a cavernous form of calcium carbonate.

erosion, they are clearly older. The cores of pinnacles in the middle and northern groups, as exposed by erosional stripping of a mantle of cavernous tufa, are composed of hard, solid stony tufa. These core-mantle relationships may indicate two stages of formation. Lake Searles tufas are not restricted to pinnacles. Small deposits form encrustations on rocks, knobby mantles on the ground, low stubby cones, and irregular mounds and ridges.

How did these huge towers and ridges form? Tufa can form by chemical precipitation from carbonate-rich waters of various origins, commonly lakes and springs, or by biochemical precipitation by microorganisms such as bacteria and algae. It forms solid stony masses and porous, cavernous, friable jumbles of intertwined calcium carbonate filaments resembling something run through a pasta-making machine.

The nature of tufa varies considerably within individual pinnacles and between the pinnacle groups. Many factors influence tufa formation, including changes in water levels, water temperature, and water chemistry. Tufa can precipitate chemically from carbonate-bearing water by temperature changes, evaporation, input of calcium-bearing groundwater, and loss of CO_2 from solution in the water. This last process may seem backward to chemists but has to do with changing acidity. Algae living in the water cause calcium carbonate to precipitate by removing carbon dioxide from the water as they photosynthesize. Most tufas contain considerable organic matter, much of it probably of algal and bacterial origin, and they also contain casts that replicate individual microbial forms.

Tufa towers in the northern group have a core of solid, rocklike tufa encased within a 10- to 20-foot-thick surficial mantle of cavernous tufa with various open textures. On some towers, part of the cavernous mantle has peeled away, exposing the core. The mantle was almost certainly deposited from waters of Tioga-age Lake Searles. The inner solid cores may have been deposited during the rising phase of the Tioga-age lake, or possibly during the preceding Tahoe stage, and were exposed during the Tioga-Tahoe interval when the lake evaporated to dryness.

Viewed from the air, the pinnacles are clearly aligned in several directions. Alignment and localization of pinnacles exist because algae prefer fresh spring water to brackish lake water, so they cluster in colonies around spring outlets. You can see this today at pinnacles growing in Mono Lake and elsewhere. At Mono Lake, springs continue to flow from the tops of some towers, making them ever higher. The floor of Lake Searles along the flank of Spangler Hills could have had many underwater springs fed by abundant water from as far away as the Sierra Nevada. Many of these springs probably lined up along fractures in the Spangler Hills granitic rock that underlies the nearshore part of the

Searles Lake floor. Such springs presumably existed when the climate was much wetter than it is today.

During the 2019 Ridgecrest earthquake sequence, a block that formed the head of a fanciful stone giant fell off the giant's shoulders and bounced down the slope. The poor headless fellow is at the north end of the northern group and is easy to reach (35.6196, -117.3720). Look around and you'll notice that few pinnacles have blocks that look as loosely held as this one was. If there is a Ridgecrest-size earthquake in the local area every two hundred years or so, that's about one hundred large shocks since Lake Searles dried up and the towers were exposed. As a result, most of the towers have had loose pieces shaken off already, forming the large loose blocks scattered around tower bases.

Under a full moon, the pinnacles cast enchanting shadows that invite us to imagine this desert landscape during wetter glacial times: an expansive lake with bubbles surfacing from underwater springs and a linear chain of tufa islands that breaks the water's surface like the ridges on a dragon's tail.

A 5-mile-long tentacle projects from the northern group along a linear fracture. Most pinnacles in this group are 10 to 50 feet tall.

VIGNETTE 17: TRONA PINNACLES OF SEARLES LAKE 145

Above, one of the stone giants in 2016. Below, the same view in 2020 after the Ridgecrest earthquake. Seismic shaking loosened the block that formed the giant's head. It gouged a deep pit in the conical debris pile, bounced down the slope, and came to rest in the foreground. The block is about 5 feet high.

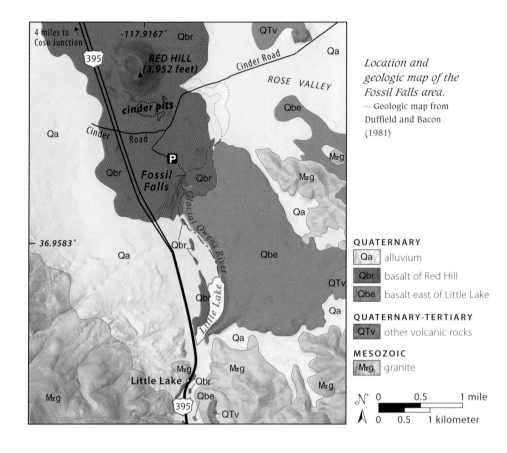

Location and geologic map of the Fossil Falls area.
—Geologic map from Duffield and Bacon (1981)

GETTING THERE

A short detour off US 395 and a brief walk bring you to spectacular Fossil Falls. Turn east on Cinder Road about 12 miles north of Pearsonville for northbound travelers and 4.9 miles south of the Coso Junction rest area for southbound travelers. Cinder Road, which is unpaved but easily passable for 2WD vehicles, takes you east along the south flank of Red Hill. Along Cinder Road 0.6 mile from US 395, turn onto the well-traveled right branch that heads south. Follow it 0.25 mile to a well-traveled side road to the east (left). Follow the side road 0.4 mile to a parking area (35.9719, -117.9109) with picnic tables and a primitive restroom. Take the rough, stony trail that leads east from near the information board. Signs of foot travel and faint orange paint on rocks mark the trail's course across the uneven lava. It is a leisurely 10- to 15-minute walk to the two-tiered dry falls, which lie about 100 feet to the right from your junction with the abandoned riverbed.

View the falls from both sides. Exercise care and common sense at the falls. The smoothed and polished rock surfaces are slippery, particularly when wet from rain, and the walls of the chasm and its potholes are vertical to overhanging. Keep children close at hand when near the falls and potholes.

VIGNETTE 18

FOSSIL FALLS
A Relict of an Ice Age River

During each of the several Pleistocene ice ages, the Sierra Nevada was heavily laden with snow and glaciers. Most of the meltwater east of the Sierran crest flowed into the Owens River, greatly increasing its discharge. During glacial times, the Owens River filled pluvial Lake Owens near Lone Pine to overflowing at a depth of 250 feet, allowing the river to continue south through a meadowed valley, now Haiwee Reservoir, and then into a wide, shallow pond in Rose Valley around Coso Junction. From Rose Valley the river cascaded over Fossil Falls, flowed south over lavas at Little Lake, and continued into Indian Wells Valley, where it created Lake China. When flow was sufficiently high, the water then flowed east into Lake Searles, then Lake Panamint, and possibly, at that lake's highest stage, into Death Valley's Lake Manly (see map in Vignette 8).

Along its course the Glacial Owens River carved the deep gorge of Fossil Falls, once a roaring cataract but now dry. You won't find fossils here; the upper and lower falls themselves are the fossils, relicts of glacial times. Significant water has not flowed over the falls for the past 15,000 years.

Let's follow the path of the Owens River upstream from Inyokern on our journey to Fossil Falls. Northbound travelers on the long, straight stretch of US 395 north of Inyokern and the US 395 and CA 14 junction will notice dark basaltic cinder cones in the Coso Range ahead and east of US 395. This is the southern part of the Coso volcanic field, a constellation of basalt cinder cones and rhyolite domes. The Coso Range hosts abundant petroglyphs and Coso Hot Springs, once a destination resort. In 1943 the Coso Range and surrounding areas were incorporated into a missile testing range, now the Naval Air Weapons Station China Lake. The area near Coso Hot Springs is a major source of geothermal power today.

About 4 miles north of Pearsonville, US 395 swings left to avoid a gently curving bluff, 0.25 to 0.5 mile east of the highway, which rises northward to a height of more than 500 feet near Little Lake. The 30-foot cliff at the bluff's top is solid lava, but old Sierra Nevada alluvial gravels, mantled by lava blocks shed from the cliff, underlie the remainder of the bluff. Geologists think "fault scarp" when they see a linear topographic feature, but this bluff is not a fault scarp; its base was trimmed by the Owens River when it was full of glacial meltwater. These lavas are about 400,000 years old and erupted from a cinder cone atop the bluff about 4 miles north of the bend in the highway.

After passing this cinder cone, the highway climbs and twists through the Little Lake gap, where the east base of the Sierra Nevada and the west edge of the Coso Range come close enough to shake hands. Water flowing south out of Owens Valley during glacial times had no place to go but through this gap. Lava from the Coso volcanic field also flowed into and through the gap from time to time, partly plugging it.

The erosional features you see today near the Little Lake gap were created by the youngest (Tioga) glacial runoff from the Sierra Nevada, which ceased about 15,000 years ago. That was the last phase of Glacial

Fossil Falls, looking northeast over the broad lower falls (center of photo) and the narrow, tortuous gorge of the upper falls. The lower falls is about 30 feet high; the upper is 70 to 80 feet high. Circle encloses two people.

Owens River. Earlier and probably larger glacial floods through the gap must have created other gorges and falls in the lavas that alluvial deposits and younger lavas later buried. The surface traces of these channels have not been identified, but arcuate topography and a strong magnetic anomaly related to a thick prism of lava identify one probable location of an earlier canyon, now lava filled, about 3 miles east of the present channel.

The gently curved cliffs behind spring-fed Little Lake were carved in 140,000-year-old basalt by rushing meltwater torrents of the Owens River. US Geological Survey scientists Wendell Duffield and Charles Bacon, who worked out the geology of the Coso volcanic field, called this unit the basalt east of Little Lake.

About 1.5 miles north of Little Lake, where the highway starts to curve left (west), a glance ahead to the right (northeast) looks up the gorge harboring Fossil Falls. Look east here to see columnar jointing in the basalt flows.

En route to Fossil Falls on Cinder Road, you pass close by the south flank of Red Hill, a young cinder cone composed of small red and black fragments of highly vesicular lava erupted from Red Hill's central vent. It has been extensively excavated for these fragments of porous lava

Red Hill cinder cone, 630 feet high, is 1.3 miles north of Fossil Falls. The rough, young basalt of Red Hill lava flow in the foreground erupted from Red Hill. Aerial view looking north across the Fossil Falls parking lot.

for use in lightweight concrete products, especially cinder blocks. The cinder quarrying was controversial because it threatened to destroy this remarkably symmetrical and scenic volcanic cone. Eventually, after much lobbying, a compromise was struck allowing mining of cinders to continue but only in such a way as to preserve the isolated cone's graceful profile as viewed from the highway.

The short hike to Fossil Falls crosses the irregular surface of a young lava flow from the Red Hill vent; these flows were named the basalt of Red Hill. Lava along the trail is full of little gas-bubble holes. The young lava flow extends as a thin veneer to and beyond the falls, but it does not make the falls; that job goes to the older, underlying flows of the basalt east of Little Lake, the same unit in the cliffs behind Little Lake.

How do we know which lava flow is which? They look rather similar from a distance and even when walking on them. However, if you inspect the lava around the parking area you will see that it contains abundant small (0.1 inch) crystals of green olivine and white plagioclase. These crystals characterize the basalt of Red Hill. At the upper falls, however,

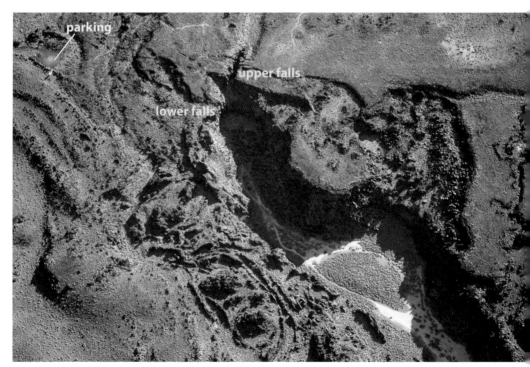

Aerial view looking down on Fossil Falls. The parking lot is in the upper left, and the trail to the falls is visible on the lava flows above the falls. The rugged, intricately eroded lava is the flow known as the basalt east of Little Lake. It is overlain by a smooth flow of the basalt of Red Hill in the upper and right sides of the photo.

Three potholes by the upper falls were drilled down a few feet by swirling sand, pebbles, and cobbles.

crystals are noticeably sparser and smaller, a characteristic of the basalt east of Little Lake. By close inspection one can tell which is which, much as an ornithologist might use subtle differences in markings to tell warblers apart.

Picture a river here deep, swift, and powerful enough to carve waterfalls and a gorge in these hard rocks. Sediment-laden water is more abrasive than clear water, and polish on water-worn surfaces suggests that the water was muddy. The falls drop in two steps about 200 feet apart. The 30-foot-high lower falls, like most falls, poured as a sheet of water over a steep edge into a plunge pool. The upper falls, 70 to 80 feet high, cascaded through a connected series of potholes.

Fast-flowing sediment-laden water carves potholes into solid rock at fixed whirlpools that are generated by high flow velocity and an irregular stream bottom. Each vortex acts like a vertical drill, increasing in size and power as it penetrates deeper into the riverbed. Once a pothole starts to form it can trap sand, pebbles, and cobbles, which then churn in the vortex and erode the pothole. As they increase in size, adjacent vortices join to create a chasm or an irregular labyrinth of connected potholes, as you see at the upper falls.

152   VIGNETTE 18: FOSSIL FALLS

Wildly irregular lava near the lip of the upper falls, cut by deep potholes and sculpted by flowing water rich in rock debris. Red pocketknife in the foreground gives some sense of scale to this surreal landscape.

Looking up to the sky from the bottom of a pothole at Fossil Falls. The length of the tube here is about 10 feet.

Rock art in the Coso Range, formed by chipping desert varnish off a flat surface of basaltic rock.

The smooth-walled complex geometry of the many integrated potholes is fascinating. With care, agile people can climb around inside this labyrinth. Climbing with ropes and other gear on the cliffs of the gorge, once popular, is now prohibited by the Bureau of Land Management. The BLM monitors and protects this area as a designated feature of Critical Environmental Concern, similar to the protection given the Trona Pinnacles (Vignette 17). Because this is a protected area, please do not remove specimens of any kind.

Before you leave Fossil Falls, find a comfortable seat along its upper bank. Study the collage of geometrical forms, then close your eyes and try to imagine the scene with the Glacial Owens River cascading by. What a delightful place Fossil Falls must have been in those times.

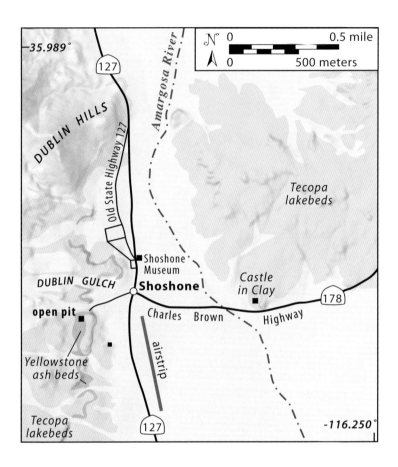

GETTING THERE

Travelers cross the Tecopa Basin between Death Valley and Baker on CA 127 and between Death Valley and Las Vegas on CA 178. Watch for good views of the Tecopa Basin as you descend the slopes 2 to 3 miles north of Ibex Pass northbound on CA 127 and also after crossing the Resting Spring Range westbound on CA 178. You can see topographic details within miniature badlands in the eroded lakebeds from both highways as you approach their intersection at Shoshone. The road through Tecopa and Tecopa Hot Springs passes through the center of the basin.

VIGNETTE 19

ANCIENT LAKE TECOPA
A Collector of Fine Vintage Volcanic Ash

The blanched landscape of fine-grained white, pale-green, and buff lakebeds in the vicinity of the little settlements of Tecopa, Tecopa Hot Springs, and Shoshone are all that remain of a series of lakes and wetlands, collectively named Lake Tecopa, that occupied this valley for about 1.8 million years during Pleistocene time when huge volcanic eruptions were occurring hundreds of miles away. These lakebeds collected a lot of volcanic ash, most of it from these supereruptions. Some of it is unusually and disturbingly thick.

The history of the lake has been worked out by Marith Reheis of the US Geological Survey and coworkers. The oldest lake sediments are about 2 million years old, deposited in wetlands and shallow lakes. The Amargosa River first entered the basin sometime before 767,000 years ago, expanding the wetlands into a large lake that then formed and lasted until about 485,000 years ago, when the river was diverted out of the basin. Then between about 480,000 and 185,000 years ago, sediments came largely from windblown and alluvial fan sources. The river returned to the basin, forming another large lake that lasted perhaps a few tens of thousands of years. Lakes, like reservoirs, can self-destruct if they overflow the dams that hold them back, and Lake Tecopa did just that by overflowing a ridge of bedrock at the south end, cutting Amargosa Canyon, flowing to Death Valley, and feeding Lake Manly.

The Amargosa drainage region covered an area more than forty times the expanse of the lake and supplied most of the lake's water. Some of it came from as far away as the lofty, often snow-capped Spring Mountains of Nevada, nearly 11,000 feet high. Springs in Ash Meadows, near Death Valley Junction, augmented the river's discharge. The climate when Lake Tecopa existed was clearly cooler and moister than it is today because the discharge of the modern Amargosa, a mere trickle, could not have maintained a lake of Tecopa's size under the present-day evaporation rate of about 6 feet per year.

The lake-floor sediments are mainly mudstone consisting of clay, silt, and fine sand, as well as shoreline conglomerate, sandstone, tufa, and numerous layers of volcanic ash. The lakebeds are slightly folded and cut by numerous high-angle faults of small displacement. Greater

Aerial view of dissected Lake Tecopa beds about 2 miles west of Tecopa. The bluffs are about 60 feet high. The Ibex Hills are in the right background.

Fine-grained, tuff-rich mudstone beds dip gently to the left (west), toward the basin center. Dark, desert-varnished gravel caps surfaces on the lake sediments. Eagle Mountain is the lone distant peak on the left, and the Resting Spring Range is on the right. View looks north across the Tecopa Basin from the outskirts of Tecopa Hot Springs.

compaction of the sediments toward the basin's center, where deposits are thickest, inclined the layers about 1 degree inward. Local settling over irregularities in the basin's floor created small compaction structures with beds tilted as much as 8 degrees.

The uppermost Tecopa beds contain sparse remains of fossil mammals such as horses, camels, mammoths, muskrats, and some rodents. Miners prospected the lakebeds for borates, nitrates, pumice, and volcanic ash. Of special geologic interest are twelve layers of volcanic ash

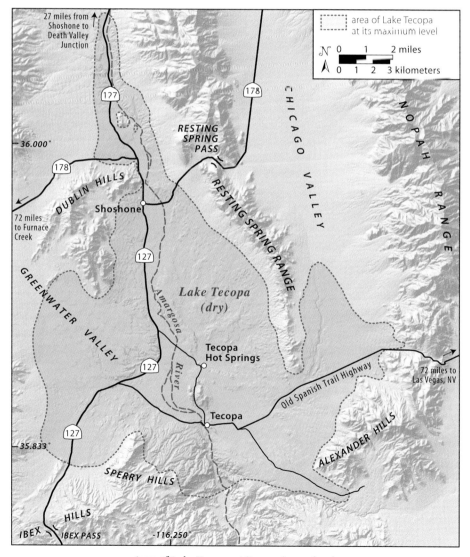

Area of Lake Tecopa at its maximum level.

within the lake deposits, three of which are particularly thick and continuous. These three layers and their sources and ages are: Lava Creek ash from Yellowstone, 631,000 years old; Bishop ash from Long Valley, 30 miles northwest of Bishop, California, 767,000 years old (Vignette 27); and Huckleberry Ridge ash from Yellowstone, a little more than 2 million years old. These were all supereruptions, far larger than any that humans have faced since civilization began.

Geologists established the source for each ash by its distinctive chemical and mineralogical composition and by determining the concentrations of a large number of trace elements such as barium, lanthanum, and thorium. These vary significantly from ash to ash and provide a sort of bar code that can be used to identify particular ashes. Ages are determined by measuring radioactive elements and their decay products (such as potassium, which decays to argon) in rocks at the eruptive center that produced the ash. Each of these three eruptions created huge calderas, and ash columns rose so high into the atmosphere that the ashes were globally distributed.

Ashes help bracket the age of Lake Tecopa. Since the Lava Creek ash is below the top of the uppermost lakebeds and the Huckleberry Ridge ash is above the bottom of the deposits, the lake must have existed in some form, but not necessarily continuously, from more than 2 million years ago to less than 631,000 years ago. Using rates at which lakebeds accumulated between the ages of the three ashes and the measured thicknesses of beds above and below the two Yellowstone ashes, a simple calculation suggests that Lake Tecopa came into existence more than 2 million years ago and lasted until about 200,000 years ago. The Lake Tecopa beds seen here are much older than the deposits from glacial-age Lake Manix and Lake Manly.

Ash layers in the lake sediments are well-exposed in several locations near Shoshone. To see the Lava Creek ash up close, go to the truck parking area on the south edge of the gas station in Shoshone. A gravel road westward from there curls around a hill into Dublin Gulch, a historic area with dozens of dwellings that were excavated in the soft ash during the 1920s. Drive about a quarter mile and park. The cliffs on both sides contain dwellings in the tuff. The basal 6 feet of this cliff is faintly layered Lava Creek ash, pure white beneath the mud washed down from overlying beds, and the upper 4 to 6 feet are ash that has been reworked by wave action in the lake and surface runoff. These reworked ash layers grade upward into typical Lake Tecopa mudstone layers. A thickness of more than 10 feet is extreme for an air-fall ash bed deposited more than 650 miles from its source, even from a supereruption. The ash here probably derives its great thickness from being washed into the lake from surrounding slopes.

Five-foot doorways lead into dwellings carved into a 6-foot layer of pure Lava Creek ash near Shoshone. Another 6 feet or so of the overlying cavernous beds are reworked ash.

Lava Creek ash consists mainly of tiny particles of clear volcanic glass. You can see them easily under a ten-power magnifier, and they glisten in the sun even to the naked eye. Most Lava Creek ash likely blew eastward because of prevailing westerly winds. Deposits are extensive in Kansas and are known as far east as Mississippi. Considering that prevailing winds blow ash eastward, how did Lava Creek ash end up as much as 700 miles southwest in southern California, where it is found in many desert lake deposits and in marine shale beds at Ventura? Computer modeling of ash plumes indicates that the great volumes of ash, pumice, and hot gases from supereruptions such as Lava Creek produce a huge umbrella cloud at altitudes more than 70,000 feet. These clouds are so energetic that they overwhelm natural winds and spread out laterally as the ash, blown to extreme heights, falls back to Earth. The original ash thickness around Lake Tecopa was probably no more than a few inches, but that's still quite impressive given how far away Yellowstone is.

The walls of an open pit quarry in this area, now closed to the public, expose thin, well-bedded layers of ash along with other layers that are contorted enough to look like dough that has been through several

160 VIGNETTE 19: ANCIENT LAKE TECOPA

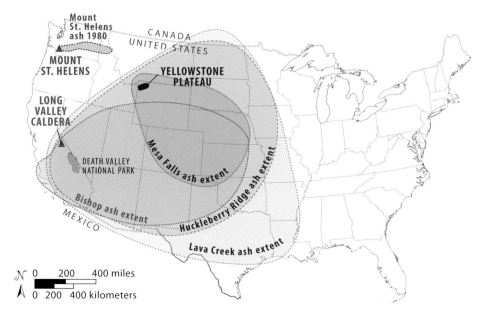

Extents of known exposures of three Yellowstone ashes, including the Lava Creek ash near Shoshone and the Bishop Tuff ash from the Long Valley caldera. Ashfall from the impressive 1980 eruption of Mt. St. Helens in Washington is dwarfed by these larger eruptions, which ejected several hundred to a few thousand times more material. —Modified from US Geological Survey

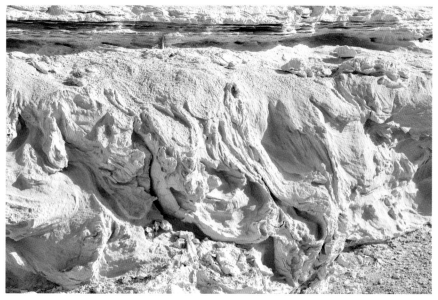

Contorted layers of Lava Creek ash, jiggled by sliding or seismic shaking while in a soupy state, are exposed in a closed quarry near Shoshone. They lie between undeformed horizontal beds of ash. The contorted layer is about 3 feet thick.

cycles of kneading. These contorted layers have flat layers above and below them, so folding clearly did not happen to the entire stack of beds. The folded beds appear to have been soft, slurpy sediments deformed at the time of large earthquakes. When water-saturated sediments are shaken hard they can liquefy, losing their cohesive strength so that denser layers above sink into them.

Highways east and south of Shoshone go past or through hills with well-exposed tuff, but it can be difficult to tell which of the three is which. You can see an excellent exposure of Bishop ash by driving east out of Shoshone on CA 178 and across the braided course of the Amargosa River. The prominent first outcrop, on the left 0.6 mile from Shoshone in a formation known as the Castle in Clay, is capped by a layer of Bishop ash several feet thick.

The Shoshone Museum in Shoshone is a must-see stop while in this area, with exhibits on local geology, Native American life, history, archaeology, and paleontology. And while you're here, you ought to head on a few miles east on CA 178 to see the welded tuff described in Vignette 20.

Pinkish-white Bishop ash layer, about 5 feet thick, caps hills of light-colored lake sediments on the east side of the Amargosa River east of Shoshone.

VIGNETTE 20

THE RESTING SPRING PASS TUFF
A Dike? A Coal Seam? No! It's Black Glass!

Those few who venture east from Shoshone on CA 178 toward Pahrump, Nevada, are treated to an unusual sight about 4 miles out of town. As the road climbs an alluvial fan and makes a sweeping turn to the right, a large southwest-facing roadcut about 0.5 mile ahead reveals a jet-black streak across its face, slanting gently (15 degrees) down to the left and coming to highway level near the cut's northwest end. For all the world, it looks like a layer or seam of coal, and one introductory geology textbook once identified it as such. Elsewhere, it has been portrayed as a dike, but it is neither coal nor dike—it is volcanic glass.

Before rushing over to lay hands on the black rock, stand back and inspect the entire roadcut and its surroundings. To the southwest is a steep mountain slope formed by massive beds of gray rock, dolostone

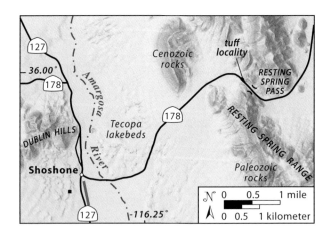

GETTING THERE

The roadcut lies 3.7 miles east of Shoshone on CA 178. The road crosses the Amargosa River a half mile east of town, traverses a beautiful landscape cut in the Tecopa lakebeds (Vignette 19) for a mile, and then begins a long climb up an alluvial fan. After a sweeping right turn 3.8 miles from town, a large parking area (35.9956, -116.2177) on the right (southwest) side of the road stands opposite the obvious bright-pink roadcut. Be particularly aware of high-speed traffic if crossing the road.

of the Cambrian-age Bonanza King Formation, about 500 million years old. In contrast, most of the roadcut is carved in a sequence of light-colored rock layers, partly sedimentary lakebeds and breccia and partly volcanic tuff, of Miocene age, tens of millions rather than hundreds of millions of years old. Many steep faults, inclined both east and west, cut these deposits with small downward displacements of the side above the fault. This arrangement indicates that the range has been stretched in a roughly east-west direction, which is consistent with the larger north-trending mountain and valleys, and indeed with the topography of most of eastern California and Nevada.

Carefully cross the highway (downhill traffic is especially fast) to where the black glass band comes to road level near the cut's northwest end. A thick, massive body of rock encases the glass layer; move a few tens of feet east up the road to see what this host rock looks like. It consists of small fragments of pumice and shards of volcanic glass along with bits of solid volcanic and sedimentary rock. This is a welded tuff, produced by rapidly moving, red-hot volcanic ash flows that partly buried the irregular surface features of an older landscape that was eroded into the even older Cambrian and Miocene rocks. Chemical analyses

The roadcut east of Shoshone as seen from the air, looking northeast. The Resting Spring Range, largely Cambrian dolostone, limestone, and sandstone, is in the middle ground, and the similar Nopah Range is in the background at the far right. These rocks are overlain in places by volcanic rocks, including the Resting Spring Pass Tuff, which forms a pale reddish-brown layer that drapes the topography and is exposed in the roadcut. The gray blob above the bright-pink roadcut right of center is a block of Bonanza King Formation dolostone that slid into the lake basin in which the tuff accumulated.

A closer view of the black glass band, which is bounded by light-colored tuff above and below. The brown discolored zone is more prominent above the band than below, presumably because gases emitted by the cooling glass rose and were more abundant there.

show that the black glass has a composition similar to rhyolite, the volcanic equivalent of granite.

Welded tuffs like this but much larger are common in the western United States. They blanketed much of Nevada and western Utah from 40 to 20 million years ago. About 18.5 million years ago, a volcano located near Kingman, Arizona, erupted the huge Peach Springs Tuff, whose ash-flow deposits stretch from the western Grand Canyon all the way to Barstow, California, a distance of more than 200 miles. Similar deposits erupted 650,000 years ago and earlier from what is now Yellowstone National Park; some of the air-fall ash from these eruptions is found nearby in the Lake Tecopa deposits (Vignette 19). The Bishop Tuff erupted east of Mammoth Lakes about 767,000 years ago and is the subject of Vignette 27. The tuff at Resting Spring Pass is less extensive, but it has many of the features of its larger cousins.

Ash flows in this area accumulated to a thickness of at least 800 feet and covered several square miles. As the hot, seething masses of rhyolite foam flowed, they broke apart. The larger pieces, golf ball size and larger, became the pumice blocks. The rest of the flow broke into tiny, sharp pieces, the volcanic ash. As the pumice and ash settled and compacted, the particles, still hot and soft, welded together firmly, forming a welded tuff.

Now go back to the road-level exposure of black glass and look carefully at its top and bottom. In both instances the contact with the host rock is a fuzzy irregular transition zone, 2 to 5 feet thick at the bottom and up to 20 feet thick at the top. These zones consist of hard, strongly welded tuff with a glassy luster and dark-brown coloration, darker in the top zone than in the bottom.

The center of the glass layer consists of relatively uniform black glass riven by an abundance of nearly vertical, closely spaced fractures. Upward, fractures are fewer, and the glass contains gently inclined thin white streaks, mostly 1 to 3 inches long. Still higher in the glass you can see scattered, flattened gas-bubble holes (vesicles). White minerals,

AIR-FALL TUFFS, ASH-FLOW TUFFS, AND WELDED TUFFS

When a volcano blasts material into the air, it comes down in one of two forms. Fine ash is lofted high into the air, typically tens of thousands of feet for moderate-size eruptions, carried downwind, and deposited in layers that blanket the countryside in the same manner as snow. The resulting deposit is called an air-fall tuff. When you see a photo of a big volcanic eruption, the eruption column consists of ash on its way up, destined for deposition downwind. Hidden inside and under the column is the other form of deposition, a frothing mass of incandescent, effervescing pumice and ash that boils out of the vent and rushes down the side of the volcano like a snow avalanche. This is an ash flow, and the resulting deposit is called an ash-flow tuff.

The foamy material in an ash flow consists of glassy pieces of pumice that range in size from fine dust to blocks as large as an SUV—all at temperatures of more than 1,300 degrees Fahrenheit. Denser than air, the glowing froth sweeps across the surrounding landscape hidden beneath the tall ash cloud that masks the destruction going on beneath. Gases leaking out of the pumice particles keep the tuff highly lubricated and mobile, making it behave like a fluid, capable of flowing around high-standing hills and leaving them as islands in a sea of tuff.

These ash flows can travel astonishing distances because exceptionally explosive eruptions can eject the material more than 70,000 feet into the air. The ash blows downwind, but the larger pumice pieces blown to such great heights pick up a lot of speed as they fall back to Earth, first downward and then outward at speeds of more than 100 miles per hour. Eventually the ash flow expends all its energy and comes to rest. It is still hot even though it cooled off during its travels. If the resulting deposit is thin or relatively cool, it simply continues to cool, forming a spongy, porous deposit of tuff. If it is thick or especially hot, residual heat softens the glass, which then compacts and welds together—hence the name *welded tuff*.

A fault (the line slanting up to the right from the first author's feet) cuts the bedding to the right (southeast) of the black glass band, which is just visible at top left. The light-pink rocks are various tuff layers. To the right of the fault, the dark rock at the base is conglomerate, and the 5-foot-thick orange layer at shoulder level is sandstone. These are the sedimentary rocks upon which the tuff was deposited. You might think that the orange rocks at shoulder level to the right of the fault line up with those high in the cut to the left, but they do not—the dark conglomerate layer is missing. This is probably a normal fault in which the left side slid down relative to the right as the crust was stretched. If so, then the conglomerate lies below the road level on the left.

mostly various forms of quartz plus feldspar and possibly a member or two of the zeolite family, thinly line the vesicles. Careful observation should convince you that the many thin white streaks in the black glass are also lined vesicles that have been completely flattened during compaction. Look higher into the lower part of the upper transition zone, and you will see many large, partly flattened gas-bubble holes, all lined with a white mineral coating.

Examine the lowermost part of the upper transition zone, and you will find some wafers of black glass, a fraction of an inch thick and an inch or so long. Their orientation parallels that of the white streaks in the glass and the compressed vesicles in the tuff. In other areas, such as exposures of the Bishop Tuff in Owens River Gorge north of Bishop (Vignette 27), pumice changes progressively from foamy pieces 1 to 2 inches across, high in the top of the deposit, to flattened black disks 100 feet deeper. During this transition the pumice pieces generally flatten without noticeably spreading out laterally, more like stepping on a foam rubber ball than on a ball made of clay.

VIGNETTE 20: THE RESTING SPRING PASS TUFF 167

Black glassy tuff with columnar jointing.

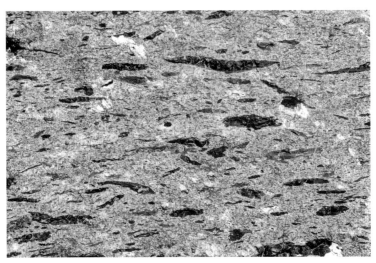
Close view of welded tuff from the lower transition zone with flattened black glassy pieces formed from pumice lumps. Width of view 3 inches.

The black color of the wafers comes from a lack of bubbles or other surfaces that reflect light. Geologists love beer analogies, and here the analogy is with a dark stout. Despite the dark-colored beer, the frothy foam on top is light-colored owing to myriad bubble walls that reflect light back to the viewer. When intensely welded, as here, pumice fragments can lose their bubbles and turn black. If welding is even more intense, as in the black band, even the broken glass pieces in the matrix become a coherent mass, have their bubbles squeezed out, and turn to black glass. The exact process is something of a mystery.

Normal volcanic glass (obsidian) forms by rapid chilling of bubble-poor rhyolitic lava. That origin is not consistent with the relationships seen here, because the interior of an ash body could not have cooled more quickly than the top or bottom. Microscopic examination of thin sections of our black glass layer reveals remnants of the fragmental texture of the original ash. Obsidian formed by chilling of molten lava displays no such textures.

Formation of this striking glass layer probably proceeded as follows. The eruption blew a pyroclastic flow of hot, gas-rich pumice over the countryside. Once deposited, the pumice fragments were hot enough (about 1,000 degrees Fahrenheit or more) that the glass particles began to compact and weld together. This process stopped early at the top and bottom of the layer because these parts cooled quickly, but the process continued in the still-hot center of the deposit until nearly all the porosity was squeezed out of the pumice, which fused into black glass. Hot gas from the crushed pumice filtered up, creating vesicles in the upper part of the glass layer and in the immediately overlying tuff, as well as depositing mineral coatings on the walls of those vesicles.

The densely welded, black glass layer is clearly not horizontal, but inclines about 15 degrees to the west. This tilting could be a result of tectonic movement of an originally horizontal layer, but the flattened pumice blocks and vesicles tell a different story. You have probably noticed that they lie at an angle to the densely welded black glass layer. Pumice compacts in response to gravity, so the little pumice disks were originally horizontal; they are now inclined eastward about 13 degrees. The cooling that controlled the location and size of the densely welded glass layer, however, was controlled by the orientations of the upper and lower surfaces of the tuff. Therefore, we can tell that the tuff was deposited on a sloping surface and later tilted by faulting or folding.

This big roadcut is a favorite stop for geological field trips because it poses questions appropriate for both students and professors. Why did such a small pyroclastic flow produce such a striking densely welded zone? That puzzle has yet to be solved.

VIGNETTE 21

ALABAMA HILLS
Hollywood's View of the Old West

The Alabama Hills, a low, scenic range of boulder piles about 9 miles long and a few miles wide, rise above Owens Valley between the Sierra Nevada on the west and the Inyo Mountains on the east. They rise only 1,500 feet above the floor of Owens Valley and aren't especially notable from US 395, but rounded granite boulders on the west side of the hills have been featured in hundreds of films and television series since the 1930s, and perhaps thousands of commercials. To many directors, the rugged boulders seem to symbolize the generic American West, and together with the high backdrop of Mt. Whitney this area has stood in for exotic locales including the Himalayas (*Gunga Din*, 1939), Afghanistan (*Iron Man*, 2008), Europe (*Gladiator*, 2000), the planet Krypton (*Superman*, 1948), and a paradise contained within a rogue energy vortex (*Star Trek: Generations*, 1994). On occasion, the area even stars as itself (*High Sierra*, 1941).

Owens Valley is as much of the story here as the lofty Sierra Nevada. Valleys typically form over time by river and stream erosion, but Owens

Aerial view of the Alabama Hills and the Sierran crest. The sediment filling the basin under the Owens River on the near (east) side of the Alabama Hills is roughly as deep as the Sierran crest is high. —USGS aerial photo, 1955

VIGNETTE 21: ALABAMA HILLS

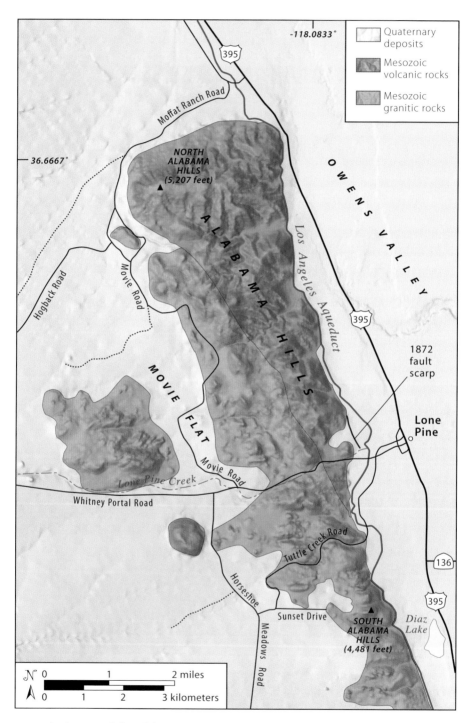

Geologic map of the Alabama Hills and locations of Movie Road and Tuttle Creek Road. Guides to movie locations are available at the Museum of Western Film History in Lone Pine.

GETTING THERE

The Alabama Hills lie along US 395 northwest of Lone Pine. Take Whitney Portal Road west from the center of town for 2.8 miles and turn north on Movie Road, which is paved for about half a mile but good graded dirt thereafter. It winds through the most scenic part of the Alabama Hills, rejoining US 395 at Moffat Ranch Road about 9.5 miles ahead. Alternatively, proceed on Whitney Portal Road another 0.4 mile past Movie Road and turn left on Horseshoe Meadows Road. In 0.7 mile you reach a dirt road on the left that takes you into an amphitheater surrounded by tall walls of jointed granite where crucial scenes of the 1939 epic *Gunga Din* were filmed.

Tuttle Creek Road is another pretty drive through the hills. To follow it, drive west from Lone Pine on Whitney Portal Road 0.5 mile and turn left 500 feet before the Los Angeles Aqueduct. In about a mile, Tuttle Creek Road crosses the aqueduct and shortly thereafter turns right to follow Tuttle Creek through the hills. Initially, outcrops along the road are drab volcanic rocks, but in about a quarter mile the road passes into beautiful granite of the sort that makes the hills famous. Upon exiting the hills, turn right on Sunset Drive and follow it 0.3 mile to Horseshoe Meadows Road. A right turn will take you back to Whitney Portal Road.

The Museum of Western Film History, located at the south end of Lone Pine, provides an excellent cultural complement to your geologic study in the Alabama Hills.

Valley is a depression formed by stretching of the crust between two major fault systems, one along the eastern base of the Sierra Nevada and the other along the western base of the Inyo Mountains. Intermittent large earthquakes on those faults may have produced only around 5 to 10 feet of vertical displacement at a time, but a thousand or so of such earthquakes during the last few million years would give us the impressive topographic relief we see today. The movement history of these faults is more complicated than just vertical displacement, though, as the Lone Pine fault (Vignette 22) illustrates.

Two main rock types are found in the Alabama Hills. One is a brown, orange-weathering volcanic rock of no particular scenic virtue, as you will see up close when walking to the 1872 fault scarp a couple miles west of Lone Pine (Vignette 22). This rock, around 150 to 200 million years old, is exposed along the first part of the drive to the mouth of Tuttle Creek and makes up much of the northern Alabama Hills. The other rock type is granite, 102 million years old, that makes up the photogenic outcrops and boulders in the Alabama Hills.

As you drive west out of Lone Pine along Whitney Portal Road, you come to the volcanic rocks (technically metavolcanic rocks, because they were heated and metamorphosed when the hot granite intruded them) about 0.7 mile from US 395, where the road and Lone Pine Creek

Typical "cowboy rocks" scenery along Horseshoe Meadows Road. Watch out for the bad guys, although here you might watch out for elephants bearing cannons because the temple scene in the 1939 movie Gunga Din *was filmed in this area. The rounded and varnished nature of Alabama Hills granite might suggest that it is much older than the pristine granite of the Sierran crest, but the two granites are nearly the same age. The Alabama Hills granite has suffered a greater degree of weathering.*

HOLLYWOOD AND THE ALABAMA HILLS

Kevin Bacon battled giant worms in *Tremors*, Clint Eastwood shot at bad guys in *Joe Kidd*, and Robert Downey Jr. demonstrated advanced weaponry in the Alabama Hills in *Iron Man*. More than 150 movies, about a dozen television series, and numerous commercials have been filmed here. Part of the appeal of the location is its proximity to the Hollywood studios, part is the dramatic backdrop of Mt. Whitney towering 10,000 feet above Lone Pine, and part is the rocks themselves: rounded granite outcrops and boulders, thousands of them, arranged so that every desperado in the West could find a good hiding place.

Movie viewers with an eye for landscape will recognize the Sierra Nevada and Alabama Hills in many films and scratch their heads at how they are portrayed. For example, in the 2000 film *Gladiator*, Russell Crowe races through a snowy pine forest, then immediately into the desert in the Alabama Hills, with Mt. Whitney featured prominently, and then from the granite desert immediately into the lush countryside of northern Italy. The geological and botanical juxtapositions are head-spinning.

squeeze through a narrow gorge. Big boulders of various types in the gorge have been tossed together in a jumble by floods and debris flows. Shortly after you exit the gorge and before Movie Road branches to the right (north), you'll find excellent outcrops of granite. Tuttle Creek Road offers equally beautiful exposures.

The rocks between the gorge and Movie Road and along Tuttle Creek are good examples of spheroidal weathering, a process that, as the name implies, weathers many types of rocks into rounded shapes and forms weathered layers like the skin of an onion. Some rock types yield better rounded stones than others. In California, the most-rounded forms develop in granite—the Alabama Hills, parts of Joshua Tree National Park, and the Granite Mountains about 70 miles east of Barstow provide good examples.

When a granite magma intrudes the crust, it is hot (about 1,400 degrees Fahrenheit) and under high pressure (perhaps 3,000 atmospheres or more). When the overlying rock erodes to expose the granite at the surface, it has cooled significantly and is at one atmosphere pressure. This dramatic change in temperature and pressure causes the granite to crack. It cracks because it shrinks as it cools, and also because of the pressure release as miles of rock above it are removed by erosion. Quartz-rich rocks like granite also contract when they cool below about 1,000 degrees Fahrenheit because at that temperature quartz changes its crystalline form to one that is about 5 percent smaller.

All these cracks, if randomly oriented, would shatter the rock. However, in many rock bodies the cracks align for mechanical reasons into sets of discrete planes called joints. The spectacular columnar joints at Devils Postpile (Vignette 31) and in Owens River Gorge (Vignette 27) are examples of joints that formed perpendicular to a cooling surface. In granite, however, three mutually perpendicular sets of joints commonly develop: two steep ones at right angles to one another and a third set nearly parallel to the Earth's surface. These flat joints are well displayed along Tioga Road in Yosemite National Park and are responsible for much of the dome topography there. In the Alabama Hills, though, vertical joints are more important, and they provide passageways for water to get into the rocks.

Given enough time to do its work, water can cause tremendous damage to rock via two mechanisms: frost wedging and chemical alteration. In frost wedging, a thin film of water gets into a crack in a rock, freezes, and expands. Most liquids contract when they freeze, but water is one of only a handful of liquids that expands when it freezes. High mountain peaks provide an ideal environment for frost wedging because the temperature crosses the freezing point daily during much of the year, and abundant water is available in the form of snow. During the day, water

Aerial view of two vertical joint sets at right angles to one another. All those cracks provide myriad pathways for water to enter and weather the rock.

seeps into and fills cracks; at night, when the temperature drops below freezing, the water freezes and expands, putting stress on the crack and wedging its icy tip a bit farther into the rock. Each daily cycle widens and lengthens the crack only a minuscule amount, but after thousands of years and many thousands of such cycles, the result is a shattered mass of rock. Consequently, many Sierran peaks are mantled with rock piles.

Chemical attack by water, though more subtle than frost wedging, can be equally devastating. Water by itself does not strongly corrode rocks, but the chemicals that water carries do. Particularly powerful are surplus hydrogen ions (H^+), which make the water slightly acidic. This dilute acid has little effect on quartz but readily attacks feldspar, the other abundant mineral in granite, turning it into clay, a weak, chalky mineral that expands and easily falls apart. Once the feldspars in a granite have turned to clay, the granite disintegrates into its component minerals. Hit it with a hammer and it will crumble or thud instead of ringing. Landscapers call the product DG—decomposed granite.

Chemical attack and disintegration account for the rounded shapes of weathered rocks. The sharp edges and corners that jointing produces are prime places for chemical attack because the edge offers two surfaces for solutions to attack and for deeper weathering that rounds off the edge. After thousands of years, the rock is well rounded. This weathering takes place mainly beneath the Earth's surface, and then erosion exposes the rounded cores of formerly angular joint blocks.

Some rocks in the Alabama Hills have been intensely desert varnished. Here, along Tuttle Creek Road, the orange-and-black rock in the center of the photo has several generations of varnish. Varnish is easily broken off or worn away when rocks are tumbled in floods and debris flows, so multiple generations of varnish indicate an interesting life. The well-rounded boulder in the foreground has darker varnish below a distinct line. The lighter upper half was probably buried when the darker lower half was being varnished, indicating that the rock was probably flipped somehow some time ago.

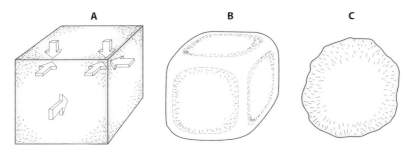

Chemical attack is faster at edges than at flat surfaces and even faster at corners. This process produces rounded rocks. (A) Stronger chemical attack at corners and edges than at faces. (B) Partially rounded block. (C) Sharp edges are gone; weathering is uniform.

This erosional process leaves us with a remaining question: Why aren't frost-shattered rocks on Sierran peaks rounded off into cowboy rocks? The answer is because the speed of the chemical reaction that transforms feldspar into clay depends strongly on temperature. The average high temperature in the mountains is far lower than in Owens Valley, so chemical reactions proceed much more slowly on the peaks. In even warmer and wetter climates, such as the southeastern United

States, the reactions run much more quickly, and fresh, unweathered granite is difficult to find near the surface.

Owens Valley, on the east side of the Alabama Hills, is remarkably deep. One way to estimate the depth of a sediment-filled basin, much cheaper than drilling, relies on gravity measurements. Basin-fill sediments are less dense than bedrock owing to pore space, so gravity over a sediment-filled basin is lower than that over bedrock. The large gravity deficit east of the Alabama Hills indicates the presence of a basin fully 10,000 feet deep and implies that the Alabama Hills are bounded by a buried escarpment as tall as the towering Sierran escarpment.

As you stroll or drive or camp in the Alabama Hills, study the granitic outcrops up close. You can pull to the side of the road just about anywhere and scramble around on the rocks. Look for rock arches. Climb high points for a great view of the Sierran escarpment to the west and try to imagine the deeply buried valley east of the Alabama Hills. With some effort you can spot the places where John Wayne, Douglas Fairbanks Jr., Russell Crowe, Kevin Bacon, and others plied their trade.

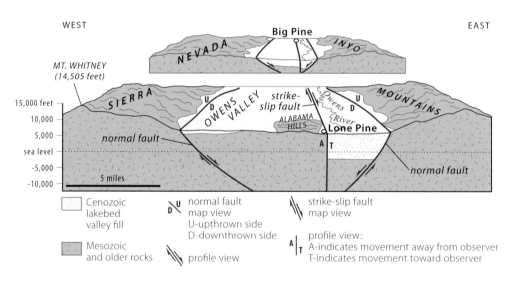

Greatly simplified view of the structure of Owens Valley, looking north. The Owens Valley fault, dominantly strike-slip, runs down the valley through Big Pine and Lone Pine. "A" and "T" indicate relative motion away and toward the observer. Dominantly normal faults bound the ranges on either side and serve to elevate them above the valley. Gravity data show that sediment filling the basin is quite thick east of the Owens Valley fault, whereas sediment fill west of the Alabama Hills, and farther north toward Big Pine, is considerably thinner.

VIGNETTE 22

LONE PINE FAULT AND THE 1872 EARTHQUAKE
Scars from One of California's Largest Historic Quakes

One great attraction of Owens Valley to geologists and cinematographers alike is the dramatic topographic relief between the valley and the Sierra Nevada. The valley floor at Lone Pine lies at about 3,700 feet above sea level, whereas the crest of Mt. Whitney, 13 miles to the west, reaches 14,505 feet. One of the greatest escarpments in the United States, this nearly 11,000-foot change in topography testifies to the tectonic processes that raise mountains and drop valleys. In 1872 a great earthquake broke the ground in this region, with some surprising revelations about how Basin and Range tectonism works. Death Valley and the neighboring Panamint Range tell a similar story, but as of this writing great earthquakes have not broken that ground in the past few hundred years.

The classic block-fault interpretation of Basin and Range geology (Vignette 7) predicts that vertical displacement on faults around Lone Pine ought to be west-side-up, consistent with the huge Sierran escarpment to the west. As we will see, however, that was not the whole story in 1872. Some faults produced dominantly dip-slip (one side up) displacement, some dominantly strike-slip (sliding), and some are a mixture of both. The 1872 event was mixed, about two-thirds strike-slip and one-third dip-slip.

Normal faults produce far more dramatic effects in the landscape than strike-slip faults. Two or three miles of dip-slip can produce a 10,000-foot fault scarp, like the ones that bound each side of Owens Valley. Two or three miles of strike-slip may have no topographic expression at all; the blocks merely slide past one another. The huge amount of strike-slip that has occurred along the San Andreas fault (160 miles or more) has produced little in the way of mountains—just wrinkles cut by a furrow that marks the path of the fault.

Let's begin our field study of the earthquake by following the directions at the beginning of this vignette and walk to the top of one of the most notable fault scarps anywhere in California. On the way you'll walk over and around cobbles and boulders of pale granite that are typical of Mt. Whitney and other peaks in this region. Boulders of this granite, transported by Lone Pine Creek, cover most of this alluvial fan. The granite is distinctive in having large crystals, up to 3 inches or more in length, of the feldspar mineral orthoclase. Granites of eastern Yosemite National Park around Tuolumne Meadows are similar in appearance and age. The Alabama Hills here also consist of various older volcanic rocks that are typically brown, rusty orange, or gray. We will use those color differences a little later in our hike to learn about the fault's displacement.

When you reach the crest of the 15- to 20-foot-high Lone Pine fault scarp (36.6060, -118.0757), try to imagine being here around 2:30 a.m. on March 26, 1872. Now most geologists are of two kinds: those who

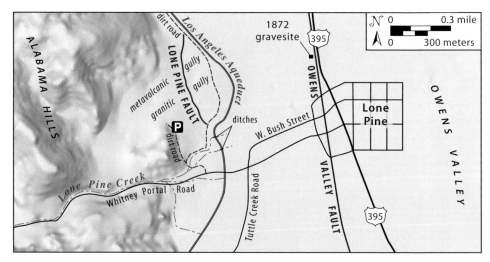

Lone Pine fault scarp location west of Lone Pine and the trace of the Owens Valley fault on the west edge of Lone Pine.

GETTING THERE

You can easily get to a zone of ground that broke and uplifted during the great 1872 Owens Valley earthquake. At the stoplight in the center of Lone Pine, turn west from US 395 onto Whitney Portal Road. Cross the Los Angeles Aqueduct 0.6 mile from the stoplight. Turn right onto the first dirt road about 100 yards past the aqueduct. Follow this road north across two rather inconspicuous tree-lined irrigation ditches and continue 0.15 mile to a broad parking area (36.6055, -118.0773) at the base of the hill on the left. Walk about 500 feet east, toward Lone Pine, to the prominent east-facing fault scarp. It is a valuable scientific resource; please take care not to disturb it.

VIGNETTE 22: LONE PINE FAULT AND THE 1872 EARTHQUAKE 179

The backlit fault scarp west of Lone Pine casts a shadow in this aerial view looking west. The Los Angeles Aqueduct crosses the middle of the image from right to left (north to south). It crosses the 1872 breakage in the boxed area and likely again near its intersection with Whitney Portal Road. The Owens Valley fault zone, which broke with 20 feet of strike-slip in 1872, is parallel to and about 100 feet west of US 395.

have never experienced a large earthquake and wish they had, and those who have experienced one and wish they hadn't. It is an eerie experience to stand on a fault scarp, especially this one, and imagine what a terrifying experience the earthquake must have been. During the earthquake everything west of the scarp jumped about 15 feet to the right and 6 feet up relative to ground on the east—all in a few seconds. The jolt would have knocked you off your feet, and you would be wondering if the shaking, which lasted 2 to 3 minutes, would ever stop.

The Lone Pine fault scarp, a branch of the great Owens Valley fault, is one of the most impressive manifestations of faulting during the 1872 earthquake—or any earthquake, for that matter. It is the highest of dozens of ground breaks in the Lone Pine area and probably formed during at least three separate earthquakes over the last 10,000 years, according to detailed studies by US Geological Survey scientists. Looking both east and west, you can see several smaller scarps cutting the surface of the alluvial fan.

Walk north along the base of the scarp toward a gully that cuts it. Notice that the fan surface east of the scarp is dotted with huge boulders,

The Lone Pine fault offset debris flow deposits consisting almost entirely of granite cobbles and boulders shed from the Sierra Nevada. The scarp height here is 20 feet, which has probably accumulated in several earthquakes over several millennia. During the 1872 earthquake everything on the west side, where the second author is standing, jumped 15 feet to the right and 6 feet up in a matter of seconds. (36.6060, -118.0757)

Aerial view looking north along the 1872 Lone Pine fault scarp, here about 20 feet high, west of Lone Pine. Mesozoic volcanic rocks of the Alabama Hills are in the upper left corner of the photo; the Inyo Mountains are in the upper right corner.

some up to 25 feet long, as big as a short school bus. These are considerably larger than the rest of the boulders in the fan, and their origin is something of a mystery. They did not fall here during the earthquake (there is no place for them to have fallen from). Most of them were probably carried here by glacial outburst floods, catastrophic events that happen when glacial ice damming a lake high in the mountains fails. Such floods may have been common during the Tioga glaciation when glaciers filled many canyons on the east side of the Sierra Nevada.

Continue walking north about 100 yards and to the north side of a deep gully where a dirt road descends from the top of the scarp to its low side. Use the gully to get back onto the crest of the fault scarp. About 100 to 150 yards north of that gully, the color of the rocks on the fan surface becomes noticeably darker, and the rocks are smaller and more angular than the granite detritus. These smaller, darker rocks occupy an old stream channel that filled with debris flows from the Alabama Hills during particularly heavy downpours. Notice that the edge of the debris flow is sharply defined by big, round, white boulders on the south and

Vertical aerial view showing brown debris flow (demarcated by red dotted lines) cut by the Lone Pine fault (yellow dotted line). The south (right) edge of the debris flow is right-laterally offset about 100 feet along the Lone Pine fault from its feeder channel. The total offset was accomplished in several large earthquakes over a few millennia rather than just during the 1872 earthquake. (36.6096, -118.0773)

182 VIGNETTE 22: LONE PINE FAULT AND THE 1872 EARTHQUAKE

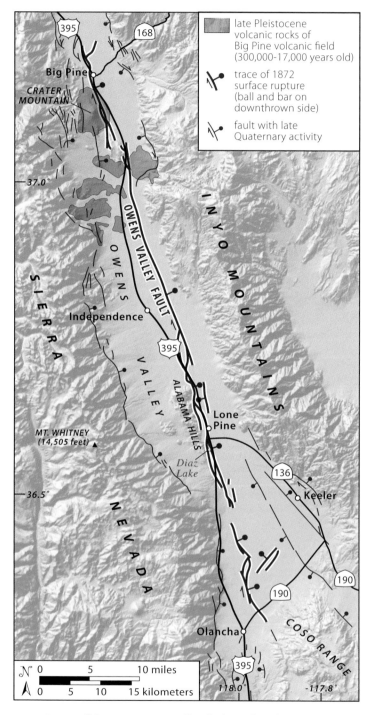

Extent of the 1872 Owens Valley fault earthquake rupture.

small, angular, dark stones on the north. Detailed mapping shows that this debris flow and channel have been offset about 100 feet right-laterally in several large earthquakes.

The Lone Pine fault scarp is certainly impressive and being here during the earthquake would have been terrifying, but the big show in 1872 occurred on the Owens Valley fault in the center of the valley. The earthquake was one of the largest shocks ever to hit California in recorded time. Its magnitude is estimated at about 7.9 judging mainly from measured fault offsets, the length of the fault that broke, contemporary newspaper accounts, the area over which it was felt, and building damage.

The newspaper accounts, quite fantastic at first but more and more reliable as more information came in, tell the story. The earthquake struck at 2:25 a.m. and leveled nearly every adobe, brick, or stone building in the Lone Pine area. Twenty-three of Lone Pine's 250 to 300 residents perished. Camp Independence, an army post 15 miles to the north, was closed after its adobe walls caved in. The quake caused rock avalanches in Yosemite Valley, 110 miles to the northwest, and was felt from Canada to Mexico, but especially over most of California and Nevada.

Word of Lone Pine's devastation was slow to reach the outside world. The *San Francisco Chronicle* noted in its March 26, 1872, edition that residents of Sacramento (220 miles northwest of Lone Pine) ran outside in their "frilled night shirts." The March 31 edition carried more news and squelched earlier rumors of giant chasms and volcanic eruptions. By the April 7 edition, however, the volcanoes were back again as reported by the newspaper: "Two volcanoes are reported in active eruption, emitting rocks and flames, roaring like furnaces. Eruptions in great numbers are reported from the Death Valley region, where the Earth is reported to have sunk several hundred feet." The quake caused severe ground breakage in and around Lone Pine and produced several notable scarps, but no volcanic eruptions or gaping chasms anywhere.

Relatively sparse field data and the complexity of the surface rupture inhibited attempts to determine the nature and amount of slip for more than a century after the earthquake. The surface rupture extends northward from Owens Lake almost to Bishop—a distance of 85 miles—and is largely a strike-slip fault. Right-slip of 20 feet has been measured across several segments of the fault. The Inyo Mountains have slipped 40 miles south relative to the Sierra Nevada along this fault over millions of years.

You may see examples of 1872 faulting along the Owens Valley fault by returning to US 395 and driving north 0.9 mile to where a roadside marker on the left (west) side of the highway marks a common grave for

The Owens Valley fault trace (dots) 3 miles east of Independence is marked by a line of vegetation, some low scarps, and ponds in this aerial view looking north. The pond in the middle of the photo is a sag pond, a common feature of strike-slip faulting. Imagine a fracture parallel to the long axis of the pond, and mentally place your hands on either side of the photo. Moving your hands in a right-slip sense (left away, right toward) will open up that fracture and produce a small basin that may fill with groundwater. A fracture deviating the other way would produce a small ridge.

victims of the earthquake. Fittingly, the site lies atop the scarp of one of the main strands of the Owens Valley fault.

The Mt. Whitney golf course, 1.8 miles along US 395 south from Whitney Portal Road, was built on 1872 fault scarps. The clubhouse is on a 20-foot-high, east-facing scarp modified here by landscaping. On hole number 5, par 3, golfers must hit up to a green on top of the scarp. On hole number 6, par 4, golfers hit off the scarp, down to the east, letting this geologic feature add yards to their tee shots.

More scarps break the surface at Diaz Lake, about 1 mile south of the golf course. One on the east side of the lake faces west and another faces east on the west side, defining the small basin occupied by the lake, which reportedly formed at the time of the 1872 earthquake.

Finally, drive past Crater Mountain near Big Pine (Vignette 23) in the late afternoon when the sun is low and east-facing scarps on the east flank of the mountain cast shadows that clearly define several fault scarps in the lava flows. These impressive scarps may have formed in earthquakes other than that in 1872.

US Geological Survey scientists estimate that large earthquakes on the Lone Pine and Owens Valley faults occur only every 3,000 to 4,000 years or so. Given such a long return interval, can residents of Owens Valley rest assured that another such earthquake will not strike in their lifetime? Probably not. Many capable faults with unknown histories are on both sides of the valley. The dramatic topography of the valley, so beloved by movie makers and geologists alike, convincingly proves that this area is geologically active, a fact that was underscored by the magnitude 5.8 Lone Pine earthquake on June 24, 2020, the largest earthquake in the Owens Valley fault zone since the nineteenth century.

Looking west from the tee at hole number 5, a 200-yard par 3, at the Mt. Whitney golf course south of Lone Pine. The hole is in the right midground on the upthrown western side of the 20-foot-high scarp, making the hole play a little long. Snow-clad Mt. Langley is on the left and Mt. Whitney just peeks through a low spot in the Alabama Hills to the right of center.

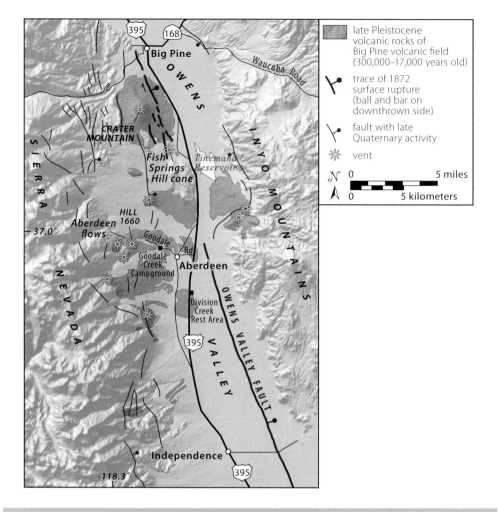

GETTING THERE

Basalt cinder cones and one small rhyolite flow are scattered across Owens Valley for 20 miles south of Big Pine. Most are on the west side of US 395. We direct you to three viewpoints, but many other places have cinder cones and lava flows up close.

To get to the Fish Springs Hill cinder cone, proceed 5.2 miles south on US 395 from its intersection with CA 168 on the north side of Big Pine. Exit at North Fish Springs Road and proceed south 1.7 miles to a dirt road on the right (west), and then 0.5 mile to the Fish Springs Hill cone.

To wander among beautiful lava-flow surface features, drive 14.1 miles south on US 395 from its intersection with CA 168 to Goodale Road; this is 5.1 miles south of the turnoff to Tinemaha Reservoir. Turn right and proceed west through the hamlet of Aberdeen (1 mile) toward Goodale Creek Campground. At 2.3 miles, 100 feet before crossing Goodale Creek and entering the campground, turn right onto a dirt road that continues for another 1.9 miles to a turn`
around (36.9940, -118.2975). This road is passable for cars with moderate clearance. There are good features in the lava flows on either side of the road, and excellent ones are found on the northwest side of the cinder cone about 2 miles to the west.

The Division Creek Rest Area, located 11 miles north of Independence and 16 miles south of Big Pine, provides easy access to lava flows. Walk south through the picnic area and across the bridge over Division Creek to a passage through the fence. The toe of a lava flow is just beyond the creek and the fence.

VIGNETTE 23

BIG PINE VOLCANIC FIELD
Lava Flows and Big Pine Cones

Young volcanism occurred in three main areas on the east side of the Sierra Nevada. Southernmost is the Coso volcanic field, home to around one hundred vents, mostly basaltic cinder cones and rhyolite domes. Northernmost are the Long Valley caldera and Inyo-Mono Craters systems, a large province near Mammoth Lakes that erupted the Bishop Tuff, Mammoth Mountain, and Obsidian Dome, all of which are covered in other vignettes. In between lies the smaller Big Pine volcanic field, a collection of about forty basaltic cinder cones and their wide-ranging lava flows.

"Young" is, of course, a relative term. Volcanism at Coso commenced about 4 million years ago and ceased a few tens of thousands of years ago, although there is faint evidence for younger activity. The Mammoth system commenced around the same time and last erupted around the year 1700, give or take several decades; clearly it is still active. Volcanism in the Big Pine field started later, around 1 million years ago, and continued until about 20,000 years ago.

The largest and northernmost vent in the Big Pine volcanic field, Crater Mountain, looms prominently over the town of Big Pine. Although the top of the cone is about 2,000 feet above the valley floor, the constructional edifice rose only about half that, as shown by numerous outcrops of granodiorite on the plateau below the cone less than 1,000 feet below the summit. The cone erupted through an elevated area of granitic bedrock. The field of geomorphology (study of the shape of the Earth's surface) is rife with unusual terms for unusual features, and we have spared you most of them, but we'll drop a fun, useful one here: *steptoe*, an isolated hill of bedrock surrounded by lava flows. Several of them poke through the flows around Crater Mountain.

Fish Springs Hill cone, the next vent south of Crater Mountain, is an easy place to explore and climb a cinder cone. Cinder quarrying and erosion have exposed the layering that underlies the surface mantle of loose debris. Cinder cones form by ejection of frothy lava into the air, which falls back and piles up around the vent. Cinders and lava bombs falling on the flank of the cone accumulate in conical layers of tephra

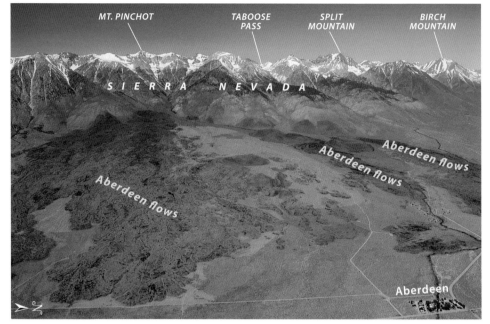

The densest concentration of vents and youngest eruptions in the Big Pine volcanic field are on the west side of the valley below Goodale Canyon. Approximately a dozen cinder cones are in this west-looking aerial photo, some fresh and some well-eroded. Extensive lava flows poured down the alluvial apron toward the Owens River. The light area between lava flows burned a few years before the photo was taken. The little community of Aberdeen is in the lower right.

Crater Mountain, 3 miles south of Big Pine, looking southwest. The Owens Valley fault scarps are the dark linear features low on the east side. Steptoes, outcrops of granodiorite bedrock surrounded by lava, are visible on the north side of the cone and hint that the cone sits upon an uplifted block of granitic basement.

Aerial view of Fish Springs Hill cinder cone looking southwest. The scar on the left side of the cone is the result of an abandoned cinder quarrying operation. The low area on the side facing the camera where the gravel road ascends the cone is a partially healed breach. The cliff in alluvium in the upper right (west of the cone) is a fault scarp along a branch of the Owens Valley fault.

that slope away from the cone, and anything falling inside the cone's rim accumulates in inward-dipping layers that are commonly destroyed by later ejections.

Before you hike up the bulldozer trail to the rim of Fish Springs Hill cone, have a look at the volcanic bombs in the quarry walls and those left behind on the quarry floor (37.0766, -118.2612). Look for sparse bits and pieces of white quartzite and glassy granite within the bombs. These rock fragments, called xenoliths ("foreign rock"), were torn off the conduit walls by, and incorporated within, the basaltic magma as it rose through the rocks beneath the volcano. The granite looks glassy and almost frothy because it was partially melted by the hot basalt.

Several of the cones in the field contain more interesting xenoliths of a dark-green, olivine-rich rock called peridotite. These are pieces of the mantle that were carried up by the magma. Geochemical analysis indicates that they came from about 30 miles deep, several miles below the base of the crust. Basaltic magma forms through the melting of peridotite, so these xenoliths provide direct samples of that source rock.

Now hike up into the crater of the Fish Springs Hill cinder cone via the gravel road on its northeast side. The loose, low-density nature of the tephra piled up around cinder cone vents makes them too weak to support rising lava, which then leaks out beneath the base of the cone.

A 20-foot-high vertical quarry wall on the east side of the cone exposes layering in the cinders and abundant bombs that blew out of the cone and fell back onto its flanks. The cone is riven with carbonate veins, although birds have done their part to whitewash the upper edge of the quarry face. Look for xenoliths in bombs on the quarry floor.

These flows typically carry away or destroy the part of the cone above them, leaving a breach that may be healed by later cinder eruptions. The road takes advantage of such a breach. Lava flows from Fish Springs Hill cone are not evident at the cone and are probably buried by alluvium.

Several vents 8 to 12 miles south of Crater Mountain built cinder cones and emitted extensive lava flows that ran about 4 miles downhill to the valley axis and presumably to the ancestral Owens River, losing as much as half a mile in elevation. A photogenic set of cones lies high on the fan surface between Taboose and Goodale Creeks. Two of these cones erupted lava, now particularly black, that flowed through breaches on their uphill sides before turning downhill and flowing down the fan.

The locality at the end of the road west of Goodale Campground provides relatively easy and isolated walking upon lava flows with many features typical of basaltic flows. Most are aa (ah-ah) lava flows, the rough kind that contrasts with the smooth, ropy surfaces of pahoehoe lava. Pahoehoe is found nearby on the northwest side of Hill 1660, about 1 mile away, and circumnavigating Hill 1660 is a worthy, challenging hiking goal. Basalt here contains crystals of fresh, clear, apple-green olivine typically up to one-tenth of an inch across, and much less abundant black pyroxene and light-gray plagioclase.

Hill 1660, in the center of this west-looking aerial view, is a cinder cone that erupted black lava out of a breach on its uphill (west) side. The cone to its left erupted a black lava flow out its north side. Both flows wrapped around Hill 1660 on their way downhill. The breach in Hill 1660 was not fully healed by later cinder eruptions, but the breach in its smaller neighboring cone was healed. The recommended excursion starts at the end of the road in lower right. The best features are found in the black lava to the right of Hill 1660. At least seven cinder cones are in the top half of this photo, many deeply eroded.

A 6-foot-high tumulus near Hill 1660 has a crusty surface of vesicular aa on its right side, but columnar joints are exposed on the left, presumably where the crust has eroded away. This feature formed as lava from the still-molten interior of the flow herniated the upper crust and leaked out, forming spatter that insulated the bulge enough for columnar joints to form.

The lava flows in this area are peppered with surface features that form when molten lava from the interior of a flow breaks through the solidifying upper crust to form a lava-filled bulge or a small secondary volcano that burps out vesicular lava like a miniature cinder cone. These widespread features are known as tumuli (the singular form is tumulus).

Mixed aa and pahoehoe lava on the flow emitted by Hill 1660. —Jorge Vazquez photo

Lavas in the Big Pine volcanic field contain abundant vesicles. In this 2-foot-high surface broken across a lava flow, the vesicles are somewhat spherical near its upper surface, but they are stretched into elongated forms deeper in the slab, indicating that flow was still molten and flowing beneath the lava's quenched surface.

Unusual worm-like lava tube on a flow surface. Beware of snakes, spiders, bats, and rodents. Hantavirus can be present in rodent nests and droppings. —Jorge Vazquez photo

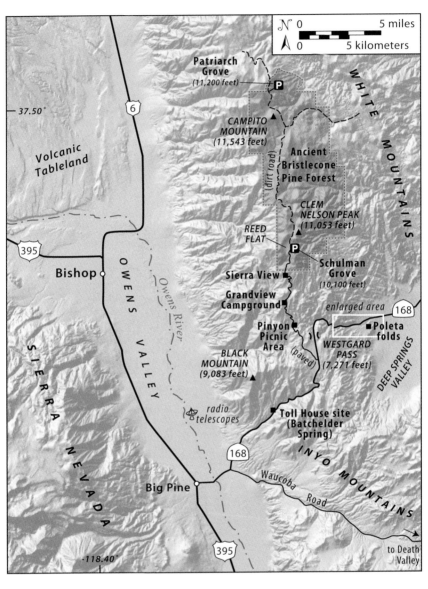
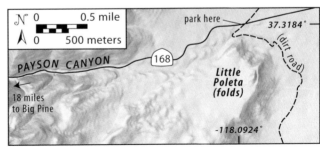

VIGNETTE 24

POLETA FOLDS
The Beds Are at Fault

Knowing the structure of the ground under your feet is important to solve problems as diverse as drilling for water, drilling for oil and gas, finding supplies of lithium and other rare elements that are critical to modern technology, tracing groundwater contaminants, and safely placing bridges and tunnels. This knowledge is the special province of geology and geophysics. Summer field camp, a four- to six-week course taught in a place with well-exposed, complicated geology, is a common requirement for Bachelor of Science degrees in geology, and for many students this capstone experience is a favorite part of their undergraduate days. For at least sixty years, more than ten thousand geology students have learned to map and interpret geologic features exposed at the surface in the Poleta folds, located on the west side of Deep Springs Valley, a little-visited area on the east side of the Inyo Mountains, east of Big Pine.

The main 1.3-square-mile mapping area of the Poleta folds lies a half mile south of CA 168, but a smaller, more accessible mapping area,

GETTING THERE

This vignette describes a 1-mile hike into the northern Poleta folds area in Deep Springs Valley east of Big Pine. Starting from the intersection of US 395 and CA 168 in Owens Valley on the north side of Big Pine, drive east over Westgard Pass about 18 miles. Once the highway leaves the narrow confines of Payson Canyon and descends into Deep Springs Valley, it makes a series of bends and then straightens out as you approach a low brown hill on the right. When the hill is at your 3:00 position (out the right-hand window) watch for a speed-limit sign and a gentle left-hand bend. At the bend, 0.1 mile past the speed-limit sign, gravel parking is on both sides of the road, and a difficult-to-see dirt road cuts back and drops steeply downhill to the southwest on the right. Park here and walk down the dirt road. If you see a dirt road veering off to the north toward a gate, you have gone a few hundred feet too far down CA 168.

Note that rattlesnakes are common in this area, and cholla cacti are abundant. Take care where you put your hands and pay attention when you sit down because you really do not want to sit on a piece of cholla—or on a rattlesnake.

Aerial view northeast over the Poleta folds area on the west side of Deep Springs Valley. Bluish-gray and light-colored bands are limestone; brown areas are sandstone and shale. All these layers were originally deposited as flat layers on the floor of an ancient ocean.

known as Little Poleta, comes almost to CA 168. The various structures are well-exposed folds and faults in Cambrian sedimentary rock layers in the sparsely vegetated, low rolling hills. In this vignette we take you on an invigorating excursion of a few hours to half a day into these folded rocks.

Geology students tramp over these hills and note the kinds of rocks, their spatial relationships to one another, and their distributions and orientations. They record their observations in notebooks and plot measurements on paper or digital topographic maps or air photos, much as an urban geographer would map streets, avenues, and buildings. But geology students go further, into the third dimension, by projecting the structures they see at the surface into the ground. And then they go further still, into the fourth dimension, by working out how these structures evolved over time. Their goal is to work out the geologic history of the area and then to add their page to the larger history of the Earth.

In the Poleta folds area, the mappable geologic features are the layers of slightly metamorphosed sedimentary rocks. These layers were deposited horizontally about 520 million years ago and probably remained that way until about 200 million years ago, when they were disturbed by thrust faults. Now the layers are inclined anywhere from gently to vertically to beyond vertically. The eroded edges of individual layers can be traced across the countryside and plotted on a map. Their inclinations are measured and recorded—information that is critical to projecting what's at the surface into the ground for 3-D reconstructions.

The Poleta folds area is ideal for teaching geologic mapping for three reasons. First, the rocks and structures in the area are well exposed and the gently rolling terrain is easy to hike around in. Second, the area is largely underlain by limestone, a soft rock that is easily deformed, leading to the folding seen here. Third, it has a distinctive and recognizable sequence of rock layers. In the area we will visit, this sequence consists of two Cambrian formations, the Poleta Formation and the overlying Harkless Formation. The study of layered rock sequences is known as *stratigraphy*, and the Poleta and Harkless Formations are but two of the many units that make up the stratigraphy of the White-Inyo Range. These units were defined by Professor Clem Nelson, who popularized this area for geologic mapping.

The rock layers in the upper part of the Poleta Formation are the key to unlocking the local geologic structure. This characteristic sequence comprises a distinctive bed of blue-gray limestone that is bounded above and below by light brownish-yellow (buff) limestone. This sequence, informally designated the buff-blue-buff unit, overlies brown sandstone and shale beds; we will call this the sandstone unit. Exposures of the

ROCK UNITS OF THE LITTLE POLETA AREA

AGE	FORMATION	GRAPHIC COLUMN	ROCK CHARACTER	THICKNESS (feet)
CAMBRIAN	Harkless		shale	section continues upward
	upper Poleta		"buff" limestone	10
			blue limestone with archaeocyathid fossils	120
EARLY			"buff" dolomitic limestone	40
	middle Poleta		shale with sandstone interbeds	section continues downward

buff-blue-buff unit are traceable on the ground and depicted relatively easily on a map, even though the beds are contorted by folding and displaced by faults in many places.

The Poleta Formation is overlain by the Harkless Formation, which is mostly dark greenish-gray, slightly metamorphosed shale. Both units were deposited in an ocean, but the sharp contact between them indicates that the depositional environment changed from the warm, shallow marine environment of the limestone to somewhat shallower tidal-flat water where mud and silt accumulated.

Ordinarily, beds of sedimentary rocks maintain a relatively constant thickness and persist laterally until they taper to a feather edge or are truncated by a fault. Faults pose a challenge to geology students; they must determine if and where truncated beds continue on the other side of the fault to determine the extent of the fault and the sense of displacement along it.

Now let's take a hike and explore the Little Poleta area. From the parking spot, walk down the dirt road to the southwest. After 100 yards the road is badly washed out for another 100 yards or so. After the washout,

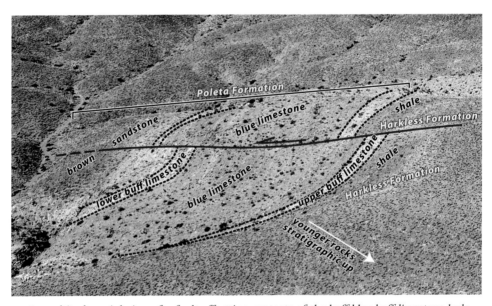

Low-altitude aerial view of a fault offsetting contacts of the buff-blue-buff limestone beds about 50 feet. Brown Poleta sandstone is visible to the left and greenish Harkless shale to the right of the limestone triplet. The fault runs through the center of the photo from left to right. In many instances fault displacement may be many tens or hundreds of feet in this area. Although you might be tempted to call this a left-lateral fault, this perspective doesn't provide the information needed to make that determination. Figuring out the sense and amount of slip on a fault is another skill that students develop through geologic mapping.

the road makes a sharp left-hand bend, climbs out of the wash, and heads east across the lower part of the large brown hill in front of you. Follow this part of the road east around the northeast side of the hill for about another 200 yards, and then walk south onto the relatively flat desert-varnished surface (Stop A). To the southwest you will see a narrow canyon on the left side of the hill, and a beautiful syncline exposed on the flank of the hill to its right.

Aerial view to the southwest outlining the recommended hike and with stops A to D from the parking spot on CA 168. The walk is about 1 mile long and involves a climb of about 200 feet.

First, admire the syncline. It is sharply outlined by a bright yellowish-white bed, which is the lower limestone of the buff-blue-buff unit. The fold is cored by the blue limestone. The upper buff limestone was removed by erosion at this location. The fold is a beauty, especially when lit by the morning sun. To its left are craggy beds of dark-brown sandstone that underlie the buff-blue-buff unit and dip under it. They're folded into the syncline, too, but the folded layers are not readily visible on the smooth slope on the right (west) side of the canyon. Note that the beds on both the north and south sides of the canyon dip to the right.

Now walk to the mouth of the canyon (Stop B) for a close look at the brown sandstone. These beds are inclined 20 to 30 degrees to the west or northwest, left to right as you look up the canyon. Note that many of the planar surfaces exposed in the wash are bedding surfaces of sediments that were deposited in a shallow sea 520 million years ago. If you break open a piece of this rock to expose a new bedding surface, you

The syncline as seen from Stop A. Left: Bright limestone of the lower buff unit is folded around the overlying (younger) blue-gray limestone. These layers overlie the older brown sandstone layers that wrap around the outside of the fold and reach to the skyline on the right side of the photo. The U shape defined by the buff limestone is about 60 yards across on the skyline. Right: A pink surface, which represents the contact of the lower buff limestone below with the blue limestone above, is projected into the air to show the broad U shape.

Ripple marks on a sandstone bedding surface in the lower part of the canyon. The grains of sand in this rock were once loose and sorted into these ripples by water currents. Shoe for scale.

are showing those quartz crystals the first rays of sunlight that they've seen in half a billion years. These layers contain patterns and structures that are created when sediment is sorted and deposited by water. These sedimentary structures are like environmental fingerprints, and among the most recognizable of these fingerprints are ripple marks that were created by water currents flowing across the sand. Geologists use ripple marks and other kinds of sedimentary structures to determine which way is stratigraphically up in a sequence of layered sedimentary rocks.

The canyon is carved entirely in the brown sandstone, and several dry waterfalls make direct ascent a tough climb. Instead, proceed up the slope on the left to the top of the canyon. The route is steep but walkable. As you ascend, you'll have a great view of a cliff cut into the buff-blue unit on the west wall (the upper buff unit is eroded away here), with brown sandstone below. Continue to the top of the canyon (Stop C), where you are standing near the contact between the brown sandstone and the lower unit of the buff-blue-buff. All these beds at this location dip to the west (left as you look down-canyon), and we know from Stop A that they flatten out and rise again in the other limb of the syncline. We would predict, then, that as we walk to the west the beds should dip back toward us.

Walk west toward the top of the hill (Stop D), paying attention to the inclination of the layering. It is harder to see on the flat top of the hill than in the canyon, but you should be able to find beds dipping back to the east, toward the canyon, on the west limb of the syncline. Can you find the layer of upper buff limestone? It is only a few feet thick in this hillslope.

View northwest of blue and buff limestone beds overlying brown sandstone in the west wall of the canyon, with ever-present geology students for scale.

It is easy to mistake surface features of an outcrop for bedding. In this view looking down at the rocks exposed at Stop D, the dark-brown planar surface in the lower center of the photo just begs to have a compass set on it to measure its orientation. However, that surface is just one side of a desert-varnished fracture surface cutting through a 10-inch-thick sandstone bed. Bedding layers in this exposure are highlighted by alternating layers of tan limestone and darker desert-varnished siltstone. The orientation of these layers is parallel to the plane of the hand and therefore quite steep. Geologists are commonly seen moving their hand in a manner similar to practitioners of Tai Chi as they work to visualize the inclination of the layering in a rock.

Before reaching the top of the hill, hold your hand parallel to the orientation of the beds in the Poleta Formation. You should be standing in brown sandstone layers that dip steeply to the east, back toward the canyon, as we saw initially when viewing the syncline from below at Stop A. This means that if you keep walking west you should remain in brown sandstone—in the parlance of field geology, you should be walking down through the stratigraphic section into older rocks. However, near the top of the hill you will encounter more buff and blue limestone. Walk west another 50 feet and you will run into green shale of the younger-still Harkless Formation. These observations do not make sense stratigraphically if the syncline continues to the west. Something must be at fault, so to speak. You will notice an overabundance of trails that thousands of students have cut through this area, a sure sign that interesting things happen here. These stratigraphic discrepancies are the sorts of complications that students must work out when learning how the Earth's crust is deformed over time.

Make your way back down to the wash by descending from the top of the hill at Stop D. You have a choice of numerous trails made by students tracing out contacts and faults. You might try to trace out the fault or faults that you just crossed. You may also encounter a number of strange things, including beds that dip opposite to one another, small folds, and sheared-out limestone that looks as if it has had a blur filter applied to it.

We will leave you with a geologic puzzle. The sandstone surface in the photo below has cryptic tracks made by some sort of creature—perhaps a proto sea slug—foraging for food in the mud 520 million years ago. Fossils such as these, which are made by creatures but do not preserve the creatures themselves, are called trace fossils. We will not tell you where to find them but will say that this spot is near the top of the brown sandstone unit below the lower buff limestone. With your newfound knowledge of the structure of Little Poleta, you should have no trouble finding them—if you haven't already.

Paleontologists seem to be quite certain that these tracks and trails were not made by students dragging their fingers through mud 520 million years ago but by a worm-like creature they named Taphrhelminthopsis.

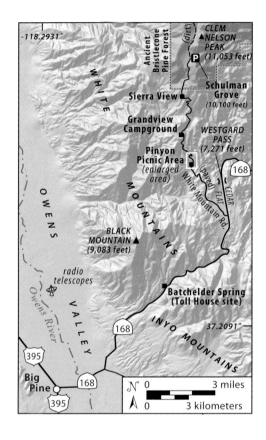
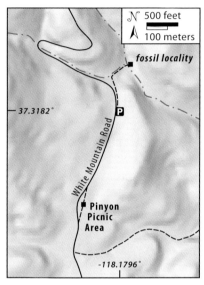

The close-up shows the location of the archaeocyathid locality in the Poleta Formation limestone.

GETTING THERE

The Ancient Bristlecone Pine Forest straddles the crest of the White Mountains, at and above 10,000 feet elevation. It's about 12 miles due east of Bishop as the hawk soars and is generally accessible from Memorial Day through the end of October. To get there, drive to the intersection of US 395 and CA 168, about 0.4 mile north of Big Pine. Take CA 168 northeast 12.6 miles to a left turn onto White Mountain Road, about 1 mile before the crest of Westgard Pass. Schulman Grove, the area described in this vignette, is 10 miles north along this paved road. We recommend two stops on the way up or down White Mountain Road. The first, a fossil stop, is near a nice picnic area 3.2 miles from the turnoff. The second, Sierra View, is at the top of a short set of switchbacks 7.7 miles from the turnoff. Sierra View is best visited in the morning when the light is at your back and shining brightly on the Sierra Nevada. Schulman Grove is 2.5 miles past Sierra View.

Bring warm clothing, sunscreen, and a hat, drink lots of water, and be prepared for cold wind, shortness of breath, and stunning, almost alien landscapes. The interpretive center at Schulman Grove has a picnic area and restrooms.

VIGNETTE 25

ANCIENT BRISTLECONE PINE FOREST
The Oldest Living Things and the Rocks They Survive On

The White Mountains are nearly as tall and steep as the Sierra Nevada, but far fewer people visit them. They are also completely different—as distinct from the Sierra as Arizona is from Maine—for several reasons. First, the White Mountains lie in the rain shadow of the Sierra Nevada and are therefore much drier. Storms that blow in from the west drop most of their moisture on the Sierra Nevada. Annual rainfall along the crest of the White Mountains is only about 12 inches, and there are no lakes and only a few perennial creeks. Second, because of the dry climate in the White Mountains, glaciers did not sculpt them in Pleistocene time. Their topography is more rounded and subdued than the jagged, ice-carved peaks of the Sierra Nevada. Third, rocks in the White Mountains are different from those in the Sierra Nevada. The highest part of the Sierra Nevada is granite—a strong, bright, bold rock, whereas the White Mountains consist mostly of somber, unassuming, ancient metasedimentary rocks of Paleozoic and late Proterozoic age, roughly 700 to 500 million years old. The two mountain ranges support different vegetation as well. Let's explore some high country to see how geology and climate affect one of the most interesting inhabitants of the White Mountains: the bristlecone pine (*Pinus longaeva*), a tree revered for its great age and noble beauty.

Bristlecone pines are the oldest known individual living plants. Some organisms, such as creosote bush and quaking aspen, grow colonies of clones that are genetically identical. Such colonies may survive more than ten thousand years, so the answer to "What is the oldest living thing?" is complicated, but we will go with bristlecone pines here. Some, more than 4,500 years old, were young seedlings when the Egyptian pyramids were constructed. In the 1950s, Edmund Schulman of the University of Arizona demonstrated their great antiquity while studying the width of annual growth rings on tree stems, which can be used to interpret past climates because wetter, warmer years generally lead to wider annual growth rings. Schulman began his studies on fast-growing species in lower-elevation forests of Arizona, but he learned that trees grown under harsh conditions, such as high elevations, commonly provide better records of climate.

Efforts to reconstruct climate in historic time led Schulman to the White Mountains. He found seventeen bristlecone pines there more than 4,000 years old and one that is about 4,600 years old. Those trees are far older than individuals of any other plant or animal species; for example, the oldest giant sequoia trees of the western Sierra seem like mere adolescents by comparison at 2,600 years old. By matching variations in tree-ring width in living trees to recently dead ones, tree-ring chronology has been confidently pushed back to the year 6827 BCE, or about 8,850 years ago. Radiocarbon dating of dead wood has helped to push the chronology back to more than 10,000 years ago, allowing scientists to compile climate records that span nearly all of Holocene time.

The drive on CA 168 to the White Mountain Road turnoff near Westgard Pass gives a good view of some of the rocks that make up the White Mountains north of the highway and Inyo Mountains south of the highway. Clem Nelson mapped the geology of this region in the 1950s and 1960s, and we owe much of our knowledge of the area to this UCLA professor. The first part of the road climbs through dull fanglomerate, and a viewpoint on the left 4.0 miles from US 395 provides a nice overview

Variations in tree-ring thickness tell a story of changing climate that goes back more than 4,000 years in living trees. By matching unique patterns from living trees with patterns from dead trees and downed wood, scientists have pushed the chronology back to more than 10,000 years before present. Some trees grow so slowly that one hundred rings make up only 1 inch of wood. How many rings can you count in this image of a sawn bristlecone pine tree trunk on the Discovery Trail in Schulman Grove?

of Owens Valley and the Palisades, a group of 14,000-foot peaks in the Sierra Nevada to the west. A few miles beyond Batchelder Spring, about 8 miles from Big Pine, the road passes through a series of narrows. In the first narrows, only one lane wide, the rocks are dark-brown to black, desert-varnished siltstone and sandstone of the Campito Formation of early Cambrian age, about 540 million years old. In the second and third narrows, the rocks are mostly pale blue and white limestone of the Poleta Formation, also early Cambrian. These rock units contain fossils of trilobites, archaeocyathids, and other species that are important to establish age and environment of deposition. We will stop and see such archaeocyathid fossils at the picnic area ahead. After the third narrows the road enters Cedar Flat, a misnamed area covered in junipers and piñon pines, with nary a cedar to be found.

CLEMENS ARVID NELSON

Clem Nelson (1918–2004) referred to the White-Inyo Range as "God's Country." He could certainly claim to know, having spent many summers there with its steep canyons, ridges, and mountain slopes underfoot while he meticulously mapped the geology of the region. The son of Swedish immigrants, Clem came to the mountains in 1948 from his student days at the University of Minnesota to study trilobites, those horseshoe-crab-like critters that inhabited the seafloor about 500 million years ago. By then a professor at UCLA, he found plenty of them, but the rocks in the ranges are so folded and faulted that if you picked up species X here and species Y there, there wasn't enough information available to tell whether X was

Clem Nelson leading a field trip in the White Mountains, circa 1993.

older or younger than Y. Clem had to work out the sequence of rock layers in the range, and to do that, he had to map the area in sufficient detail to account for the folds and faults. With support from the US Geological Survey, he ended up making and publishing several classic geologic maps in the 1960s and 1970s. These maps cover about 1,000 square miles of previously unmapped and little-known country and inspired hundreds of other geologic studies in a wide variety of fields.

Clem was a superb field geologist and a dedicated and inspiring teacher to both his students and colleagues, and he was well known for his joke-telling abilities around the campfire. The Poleta folds area in Deep Springs Valley was his preferred location to teach fledgling students the art and methods of making a geologic map. The intricacies of the area's geology inevitably attracted multitudes of students from other colleges and universities.

To honor and recognize his manifold contributions to geological science, education, and general knowledge of the White-Inyo Range and the Ancient Bristlecone Pine Forest, an 11,053-foot peak on the crest of the White Mountains 13 miles east of Bishop was named Clem Nelson Peak in 2012. The peak, carved in Reed Formation dolostone, is festooned with a thriving forest of bristlecone pine trees and overlooks the entirety of the region he mapped and loved. The hike to its top from Schulman Grove is steep, but the summit provides a remarkable view of nearly the entire area covered by this book.

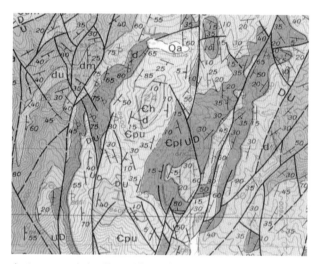

Part of a geologic map made by Clem Nelson, showing an area about 3 miles wide north of Cedar Flat. Colors are different formations, heavy lines are faults, and T symbols are measurements that show which way the rocks are dipping. The mapmaker walked to all those measurement spots, in steep terrain, to make a set of geologic maps covering around 1,100 square miles. That's an area comparable in size to a strip enclosed by Santa Monica, San Bernardino, Riverside, and Long Beach, or that enclosed by San Francisco and Vallejo down to San Jose and Santa Cruz. Did we mention that it is steep and at high altitude? —From Nelson, 1966

White Mountain Road, north from Cedar Flat and the Westgard Pass area, traverses in and out of the Campito and Poleta Formations, and you will probably be able to spot the transitions by the changes in the rock colors. The road passes Pinyon Picnic Area 3.2 miles from the turnoff from CA 168, and blue limestone of the Poleta Formation limestone nearby contains archaeocyathids, enigmatic solitary organisms that built reefs. If you would like to find some, drive north a quarter mile along the straight stretch of road north of the picnic area. Park at a gravel pullout on the right (east) side that is just before the road bends left. Walk along the right shoulder 100 yards to a faint trail that descends along yellow-green shale to a gravel wash that runs east-west. Turn right and follow this wash 100 yards to a north-south wash cut into blue limestone. This limestone contains abundant archaeocyathid fossils, and once you find a few, you'll find many more. Look parallel to layering rather than down onto bedding surfaces. They show up well around orange silty layers.

After passing Grandview Campground, the road climbs a short set of switchbacks and arrives at Sierra View. The short walk to the farther viewpoint is worth it because it provides an unobstructed view of 130 miles of the Sierran escarpment. From the viewpoint, the mighty

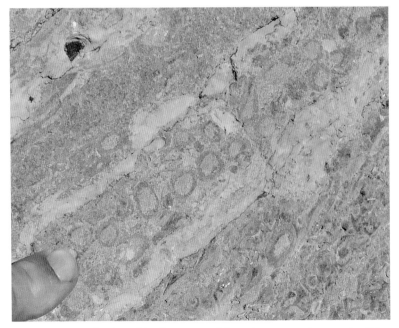

Archaeocyathid fossils in blue limestone of the Poleta Formation, north of Pinyon Picnic Area. Most of the fossils are cut nearly circular, but a few are almost parallel to the outcrop surface and are much more elongate.

CAMBRIAN LIFE

A trilobite of the genius Olenellus, *common in the White Mountains. Specimen is about 2 inches long.* —Dwergenpaartje, CC SA 3.0; image not copyrighted as part of this book

Basic animal groups originated as far back as 700 million to 800 million years ago, whereas the first undisputed fossils of animals appear only 580 million years ago. Finally, in early Cambrian time, about 540 million years ago, life-forms began to make mineralized hard shells and parts that could be fossilized and preserved.

Trilobites are foremost among early Cambrian life-forms we find fossilized. They suddenly appeared in the fossil record complete with eyes and legs. They were complex, highly evolved creatures resembling today's horseshoe crabs. Their exoskeletons were made of chitin, the same biodegradable material we find today in fish scales and the exoskeletons of insects. Trilobites and soft-bodied, worm-like forms cruised through the mud of shallow seafloors, leaving tracks and trails of their passing. Other life-forms left vertical, pencil-size tubular structures.

VIGNETTE 25: ANCIENT BRISTLECONE PINE FOREST   211

A guess as to what sponge-like archaeocyathids of the White Mountains type looked like.
—Modified from Fenton and Fenton, 1958

Palisades group of peaks, several more than 14,000 feet, dominates the left side; peaks in the Bishop Creek drainage occupy the center; and jagged Mt. Humphreys and the red-topped pyramid of Mt. Tom are on the right side. Mt. Dana, which lies just south of Tioga Pass and west of Mono Lake, is at the far right end. The green patch on the valley floor below Mt. Tom is the western part of Bishop.

The road enters an area of scattered bristlecone and limber pines less than 1 mile past Sierra View. About 1.5 miles past Sierra View, the road rounds a bend and comes to Reed Flat, a treeless expanse high on the crest of the range at about 10,000 feet. Just to the right (east) of the flat is a dense grove of bristlecone pines called Schulman Grove. The contrast between treeless Reed Flat and the relatively densely forested

This view of the Sierra Nevada from Sierra View parking encompasses a 90-mile stretch of the range, and if you walk to the farther viewpoint, you can see Mt. Whitney and an unbroken 130-mile stretch.
—Bruce Bilodeau photo

grove is curious and geologically significant. It seems that the bristlecone pines are picky about which rocks they will grow on.

Park at the interpretive center for the trailheads of two self-guided hikes, the Discovery Trail and the Methuselah Trail, that will lead you among the trees and their rocky soils. Let's walk the Discovery Trail, the shorter of the two. Less than 1 mile long, it goes in and out of the grove, from rocks that are hospitable to bristlecone pines to ones that are covered with sagebrush instead of bristlecones. Our narrative discusses the geology, and you can follow the botany with the interpretive pamphlets and signs.

Limber pines (*Pinus flexilis*) grow here in addition to bristlecone pines. The two pines are easy to tell apart, especially by their cones. Although both have needles in groups of five, bristlecone needles are darker and grow in whorls all around and along the entire length of stems, like a bottlebrush, whereas limber pine needles are lighter green

Aerial view looking to the north at Clem Nelson Peak (37.3951, -118.1731, elevation 11,053 feet), Reed Flat (left), and the Schulman Grove interpretive area (at the end of the spur road below the peak). Note the abundant bristlecone pines on Reed Formation dolostone (white soil) and no trees on the brown hill in the left middle ground (sandstone of the Deep Spring Formation). A fault separates white Reed dolostone on the right from brown sandstone of the middle Deep Spring Formation on the left. The Discovery Trail climbs through the trees at center before reaching its high point and emerging onto treeless brown sandstone. On these south-facing slopes, bristlecone pine trees are almost completely restricted to dolostone. The rough dirt section of White Mountain Road wraps around the peak, continues north, and points right at Blanco Mountain, another bristlecone-covered block of Reed Formation dolostone.

and cluster at the ends of stems. Bristlecone cones grow at the ends of branches and have fragile spines that give the tree its name, whereas the cones of limber pine are stout and do not have spines. To calibrate your eye, the tree within the parking circle in front of the visitor center is a limber pine; the trees along the boardwalk to the visitor center are bristlecone pines. Look at the trees around here and you will see that most are bristlecone pines.

Some gnarly bristlecone pine trees have only a small living segment attached to deadwood, a condition called strip bark.

As you start down the Discovery Trail, notice that the rocks here are different from those we saw along White Mountain Road. These rocks are brilliant white and weather to a light-brown soil, much lighter than other soils around here. Freshly broken rock surfaces are bluish gray. The Reed Formation is latest Proterozoic in age, about 570 million years old—older than the Campito and Poleta Formations—and consists mostly of the calcium-magnesium carbonate mineral dolomite, whose chemical formula is $CaMg(CO_3)_2$. Dolomite is like the more common carbonate calcite ($CaCO_3$) except that half of the calcium atoms have been replaced by magnesium. The Reed Formation is the only abundant dolostone formation in the White Mountains, and as you walk around the Discovery Trail you will see that the trees growing on it are almost all bristlecone pines. Why is that?

In a perverse way, the bristlecones prevail here because the growing conditions are tolerable for them but ghastly for other trees and plants. Low rainfall, cold temperatures (annual average a little above freezing), and persistent wind (average speed about 15 miles per hour) create an environment harsh enough to defeat many other plants. In addition, the Reed Formation makes terrible soil that is alkaline and deficient in important mineral nutrients, including phosphorus. Bristlecone pines can tolerate these conditions, although their growth is slow. They grow faster on other soils, such as those developed on the Campito Formation, but sagebrush grows well there, too, and outcompetes the bristlecone seedlings. The key appears to be that sagebrush cannot make a go of it on the dolomitic soil. That frees up water and light for the bristlecone saplings.

It is commonly said that bristlecone pines grow only on dolostone. This is easily disproved simply by looking at the north-facing slopes behind the visitor center. On north-facing slopes bristlecones grow on all the local rock units, including sandstone, limestone, and granite, but only bristlecones thrive, if you can call it that, on south-facing dolostone soils.

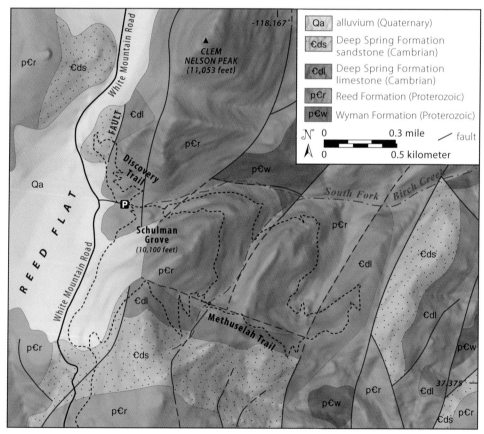

Geologic map of the Schulman Grove area, showing Discovery and Methuselah Trails.
—Modified from C. A. Nelson, 1966

VIGNETTE 25: ANCIENT BRISTLECONE PINE FOREST   215

The oldest bristlecones are not the healthiest-looking trees in the grove, but rather are gnarled, windswept, and mostly dead. Young trees typically have a full set of branches, but the older trees have spiky, dead tops, only a few living branches, and a thin strip of bark twisting around the tree to the living branches. They grow exceptionally slowly, sometimes adding only half an inch of girth per century under the thin bark strip. Harsh conditions also add to the longevity of dead bristlecone pine

Well-spaced trees, all bristlecone pines, on a south-facing hill of dolostone (upper). Bristlecone pines prefer to grow where lack of moisture and essential nutrients suppress their competitors and thus there is little underbrush. The tree spacing and lack of underbrush also keep the incidence of fire low, contributing to tree longevity. This north-facing slope (lower) is covered with a dense growth of bristlecone pines and limber pines. North-facing slopes undergo slower evaporation and keep snow for days or weeks longer than south-facing ones.

wood. Healthy, fast-growing trees tend to develop heart rot before they reach a few feet in diameter, but trees on the Reed dolostone have sound wood throughout.

Continue your walk along the Discovery Trail as it passes bristlecones in many different stages of growth, most growing on Reed dolostone. The trail climbs gently uphill through a dense stand of bristlecone pine trees, with many benches so that you may sit and re-oxygenate your body. At its high point the trail abruptly emerges from the grove onto barren slopes of angular brown rocks. Sandstone of the Deep Spring Formation, a slightly younger (about 550 million years old) rock unit, is faulted against the Reed dolostone here. Bristlecone pines are almost completely absent from this sandstone, whereas sagebrush seems more at home. Look north and south of the grove and you will see more of the same—bristlecone pines on white Reed Formation soil, only scrubby sagebrush on other rocks.

Eventually the Discovery Trail turns downhill, switchbacking down a series of steps made of brown sandstone toward a small clump of bristlecone pines that look to be growing on Deep Spring sandstone. Closer inspection, however, shows that these trees, too, are growing on carbonate rocks—in this case, limestone within the Deep Spring Formation. A general correlation between bristlecones and carbonate rocks is holding up pretty well.

Take your time studying the bristlecone pines and their favorite rocks. They aren't going away anytime soon. The older ones have stood atop the White Mountains for more than forty centuries and have seen over a million and a half days.

VIGNETTE 26

BUTTERMILK BOULDERS
A Different Kind of Highball

Rock climbing may bring to mind images of a sweaty person tied into a rope high off the ground and pounding pitons into a cliff, or a graceful athlete tied into a rope high on a cliff doing the splits to reach a distant toehold. In recent decades, though, a less elevated form of climbing—bouldering—has grown in popularity. Bouldering is what most of us did as kids, climbing up rocks and sitting upon them just because we could. The Bishop area hosts two of the best bouldering sites in the world: the Buttermilk Boulders here and the Happy Boulders about 9 miles north east as the raven flies. We will begin by visiting Grandpa Peabody and Grandma Peabody, the two huge boulders closest to the road.

In its simplest form bouldering is just climbing up on top of boulders, but like other forms of rock climbing it has evolved into harder and higher routes. Climbing typically involves rehearsing sequences of moves designed to get past a particularly difficult section. Bouldering

A climber on Green Wall Arête *on the Green Wall boulder.* —Photo by Bob Harrington

may be done by traversing just a few feet off the ground, or on terrifying "highball" routes that may rise 50 feet off the ground. Instead of being tied to a rope, bouldering relies on low elevation, friendly spotters to catch your fall, and crash pads for safety. If you see someone hike by with what looks like a folded single bed on their back, they're not going camping, they're going bouldering.

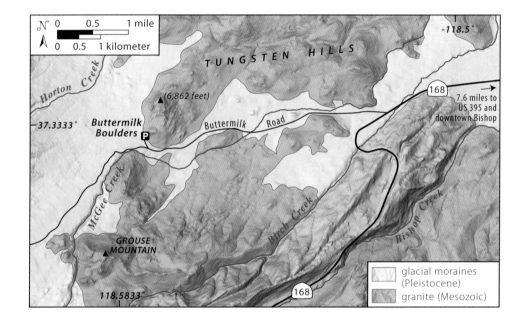

GETTING THERE

The Buttermilk Boulders are 11 miles west of Bishop in the upper Tungsten Hills. Drive west out of Bishop on CA 168 (West Line Street). The road swings left after 3 miles, and Buttermilk Road (graded dirt, washboarded, and somewhat rocky but passable for passenger cars) splits off to the right 4.3 miles later (7.3 miles from Bishop). Drive 3.6 miles on dirt to a parking area (37.3275, -118.5756) with signage near a couple of monsters known as Grandma and Grandpa Peabody.

Below we describe a short hike among the boulders, but you may enjoy the area and learn about the origin of the boulders wherever your wandering takes you. This vignette makes a good half-day trip; bring a picnic, wander among the granite boulders and bedrock, enjoy the views, and watch climbers attempt difficult routes. This area is under intense stress because it receives so much foot traffic and camping. Please stay on marked trails and granite where possible to avoid making ant-farm paths across the countryside.

Your first thought upon visiting this area might be "Wow, those are big rocks!" Indeed they are. Grandpa Peabody is about 50 feet high and 65 feet in diameter. Its scale is somewhat difficult to appreciate unless there are people near it or on a face because from a distance it looks just like another boulder.

Your second thought might be "How did these get here?" Climbing guides commonly call these boulders erratics, short for glacial erratics, which are boulders that were carried some distance by glacial ice and then deposited where the ice melted. That is certainly a reasonable interpretation given the glacial history of this region and the presence of moraines nearby, including about 2 miles south of the boulder area. A related glacial mechanism that can deposit huge boulders like the Peabodys is a monster flood caused by the failure of a glacial ice dam in the mountains. Let's lump these two mechanisms as the glacial hypothesis.

In contrast, many large granite boulders are simply cores of solid, unweathered rock left behind when weathering rounds off the corners of jointed blocks, producing rounded granite forms sitting among decomposed granite debris. These are known as core stones, and much of the scenery in the Alabama Hills (Vignette 21) consists of core stones. We will call this the core stone hypothesis.

Another possible origin is that the boulders simply rolled down from bedrock outcrops on the ridge crest above, perhaps dislodged during the large earthquakes that rock the eastern Sierra. The large boulders sit on and at the base of a significant slope that rises more than 500 feet to the crest of granite bedrock up the hill above Grandpa Peabody. Let's call this the rockfall hypothesis.

These three hypotheses can be tested by various means. The glacial hypothesis requires that the boulders were brought in from the west by ice or glacial floods flowing downhill toward Bishop, and the second two require that the rock in the boulders matches that of the surrounding bedrock. Let's take a look and sort this out.

From the signage next to the relatively small, appropriately named Whale boulder, walk northeast about 300 feet along the trail to the huge Grandma and Grandpa Peabody boulders. You will note that the climber trail is marked by chunks of bright-white quartz. Take a look at the rock that makes up the Peabodys. You will have to search around for a face that is not obscured by staining, lichen, armor plating (see pages 225 and 226), or climber's chalk. The rock is a typical granite, with roughly equal amounts of clear, smoky gray quartz, white plagioclase, pinkish-gray orthoclase, and a sprinkling of black minerals, chiefly biotite and hornblende. Keep this in mind when we find samples of undisputed bedrock.

All the boulders on this slope are made of this same granite. Now look at the mountains to the west, where you will see that darker rocks,

220 VIGNETTE 26: BUTTERMILK BOULDERS

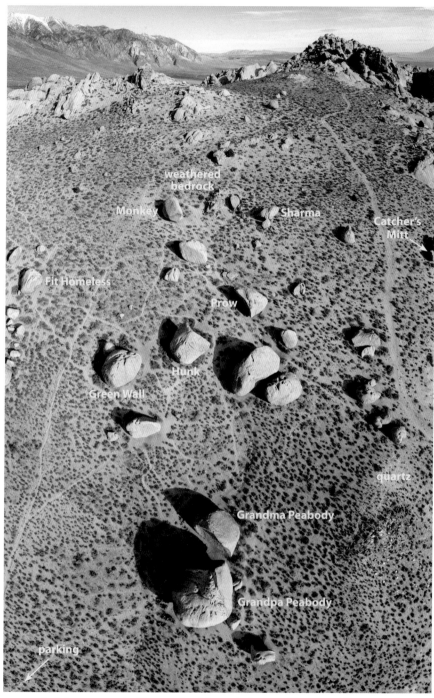

Aerial panorama looking north of the main Buttermilk area, with names of selected boulders. In this image the view to the horizon is looking horizontally, whereas the view to the Peabody boulders is looking straight down. Bedrock in upper right is about 500 feet above the Peabody boulders.

The huge Peabody boulders with sunlit Drifter in left background dwarf the person at left. The shadowed, wildly overhanging south side facing the camera is tilted fully 45 degrees past vertical, and some routes on it, such as The Process, *are so difficult that only a few people have ever completed them. The right side hosts* Ambrosia, *a highball route that takes the climber 50 feet up on a near-vertical face.*

Typical granite of the Buttermilk area. All the boulders and nearly all the bedrock is made of this rock.

in shades of dark gray, brown, and red, are abundant. The massive pyramid of Mt. Tom is particularly rich in these diorites, metavolcanic rocks, and metasedimentary rocks. Geologic maps of the eastern slope of the Sierra Nevada here show that more than 50 percent of the potential source area for the boulders consists of rocks that are not granite. This observation is difficult to reconcile with the glacial hypothesis, given that the Buttermilk Boulders are all granitic.

VIGNETTE 26: BUTTERMILK BOULDERS

Mt. Tom, whose summit is about 5 miles west of the climbing area, is largely dark diorite and reddish-brown metamorphic rocks. In contrast, boulders in the climbing area are all granite. The lack of nongranite boulders shows that glaciers flowing east off the mountains did not deposit the Buttermilk Boulders as glacial erratics.

When viewed from above, the boulders reveal a secret not visible from below—their tops are dimpled with dozens of weathering pits up to several feet across. It is also clear from this view that the two Peabodys were once a single monster boulder, about 50 feet tall, 65 feet wide, and 90 feet long. That is about 20,000 tons of rock, or about 600 dump truck loads.

From the Peabody boulders continue northeast across the slope toward a dirt road that is now closed to vehicles. In about 200 feet you will come across an area of angular quartz rocks, weathered debris from a seam of quartz in the granite. Some of the quartz is stained dark red and brown, with dark reddish-brown cubes embedded in the quartz and cubic holes where such material has weathered out. These were once pyrite crystals, and prospectors are drawn to such rocks because gold is locally associated with them. Most outcrops of iron-stained quartz are marked by at least a little futile pick-and-shovel prospecting.

Walk north on the dirt road about 300 feet to a boulder on the east side of the road whose upper surface consists of coalescing weathering pits up to 6 feet across and several feet deep. We will informally call this rock the Catcher's Mitt. These pits are equivalent to those on top of Grandpa Peabody but much more accessible. A large pit on its west side is breached low enough that only modest bouldering skills are needed to scramble up and examine the pits. The rock is the same granite we saw at the Peabodys.

This aerial view shows that the top of Grandpa Peabody (lower) is dotted with dozens of weathering pits up to several feet across. Grandma has pits, too, but they are on the northwest upper side. However, if Grandma is placed back on the fracture that separates her from Grandpa, then the pits are on top, suggesting that they formed before the fracture opened.

The pits form by frost wedging, microbial action, and other rock weathering processes. Most are closed depressions that hold rain and snow, and breached pits on the sides of boulders typically have dark, locally indented trails of lichen where water has overflowed. Top surfaces of the bedrock a few hundred feet to the north have abundant pits that are visible from higher vantage points, consistent with the idea that the pits form on surfaces that hold rain and snow.

Now circle around to the north side of the Catcher's Mitt and you will see that the rock immediately to its east has pits as well—but on the bottom side. The Catcher's Mitt has some on the north side that run into the rock more or less horizontally, and some sort of creature uses these for a home. Complete your circumnavigation of these boulders and note that their undersides are deeply undercut. Rock under the boulders is

The Catcher's Mitt boulder, with deep, huge weathering pits. The large pit in which the first author is standing, and the smaller one on the right front side, are deeper than the others because they have drained out the side, allowing accumulated debris to wash out.

North side of Catcher's Mitt (right) and its companion, which has a well-developed set of weathering pits on its underside in addition to a small number on its top. Curious.

granite, but it looks weathered and broken-up. Let's file these odd observations away for now.

The hills to the north are clearly in-place bedrock, deeply jointed in a manner similar to that in the Alabama Hills. They provide beautiful places to walk, with expansive views, hidden valleys, granite arches, and abundant opportunities for low-risk scrambling. The granite is the same granite we've seen in the boulders.

After examining the bedrock, walk to Sharma, a boulder 250 feet west of the Catcher's Mitt. Like the Mitt, it has a deeply undercut, smoothly curving underside, and you can see all the way from one side of it to the other. Sharma sits upon deeply weathered granite that looks somewhat crushed. Nearby bedrock is similarly deeply weathered and streaked with rust marks, unlike the granite in the boulders.

From Sharma work your way due south, toward the Peabody boulders, past Plow, then angle to the right a bit, past Hunk, to Green Wall. This boulder, so named for patches of fluorescent yellow-green lichen on its north side, is a good place to see boulder armoring up close. The planar north-facing wall is coated with a tough surficial layer up to about an inch thick that has cracked into polygonal plates. Edges of these plates provide climbing holds on many of the boulders. Some of the toughest climbing involves holding onto thin plates only 0.25 inch or less in thickness on overhanging slabs.

A boulder investigator peers across the smooth underside of Sharma, which has only a small contact area with the weathered granite below.

226   VIGNETTE 26: BUTTERMILK BOULDERS

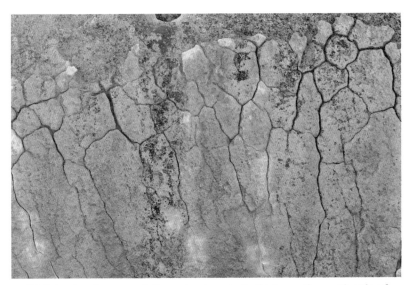

Well-developed armor plating, about an inch thick, on the north side of Green Wall. The route up the center (finger holds highlighted by white chalk) is the difficult Green Wall Center. *Width of view is about 10 feet.*

Armoring is widespread on granite landscapes, but its origin is only partially understood. Relentless attack by rain, sun, organic chemicals from living things, and other processes, collectively lumped as weathering, dissolve some components of the rock. These can be redeposited slightly deeper, and if enough are deposited to fill cracks and pores there, then that subsurface layer is somewhat protected from further weathering. Removal of the leached surface layer can then expose the hardened layer. On the boulders in this area there is a competition between weathering, which peels away surface coatings to expose rough, coarse granite surfaces, and armoring, which protects the surface. Armoring seems to be somewhat more widespread on north-facing slopes than on others, although it is present on all surfaces. See if you can detect patterns in those places where you find it.

After pondering the turtle-shell appearance of Green Wall, walk a few hundred feet northwest to Fit Homeless and find the adjoining boulder shown below. This odd boulder has several pits on its upper surface, but its lower surface is deeply pitted and looks similar to an upside-down Catcher's Mitt.

Now let's evaluate the hypotheses presented at the start of this vignette. Here are some relevant observations:

- All the boulders we've seen appear to be made of the same coarse-grained granite.
- The nearby bedrock is made of the same granite.

- Bedrock cropping out among the boulders is deeply weathered and iron stained.
- Bedrock granite on the ridge crest is fresh and well jointed.
- Much of the ridge-crest bedrock has deep pits on flat upper surfaces, but only shallow pits on sides
- Most of the boulders in this area have deep weathering pits on their upper surfaces.
- A small number of the boulders have deep pits on their sides or undersides.
- Many of the boulders, such as Sharma, Monkey, and Catcher's Mitt, have smooth, deeply undercut undersides.
- Granite under these deeply undercut boulders is highly weathered and locally smashed.

These observations rule out the glacial hypothesis because none of the boulders seems to be exotic to the area; instead, all appear to be made of the same granite as the underlying bedrock. Deep undercuts, the contrast between fresh granite in the boulders and deeply weathered granite under them, and the presence of deep weathering pits on the undersides of some boulders would seem to conflict with the core stone hypothesis. Such features are not found in the Alabama Hills and other similar granite exposures where the rocks are clearly in place.

That leaves us with the rockfall hypothesis as the leading candidate. We propose that the Buttermilk Boulders originated as jointed blocks in the ridge crest and were shaken loose during large earthquakes, rolling and sliding downslope before coming to rest in this boulder field. Some of the blocks had weathering pits on their upper surfaces, and boulders such as the one next to Fit Homeless came to rest with these upper surfaces pointing downward and have since developed pits on their new upper surfaces. A rolling/sliding origin explains both the smoothly curving undersides of many of the boulders and the smashed character of the bedrock below them.

Microscopic examination of the gray material under Catcher's Mitt confirms that the rocks under at least some of the boulders were crushed into microbreccia, similar to that produced when rock is crushed by movement along a fault. Here the work was done by sliding as the huge boulders came to rest.

Close the loop by walking back to the Peabody boulders and examining the underside of Grandpa Peabody from the cleft between the two. You will see the same story—a deeply undercut underside with smashed, weathered granite beneath. In rockfalls the largest rocks carry the most energy, so the bigger the boulder, the farther it will roll or slide downslope, and that seems to be the case here.

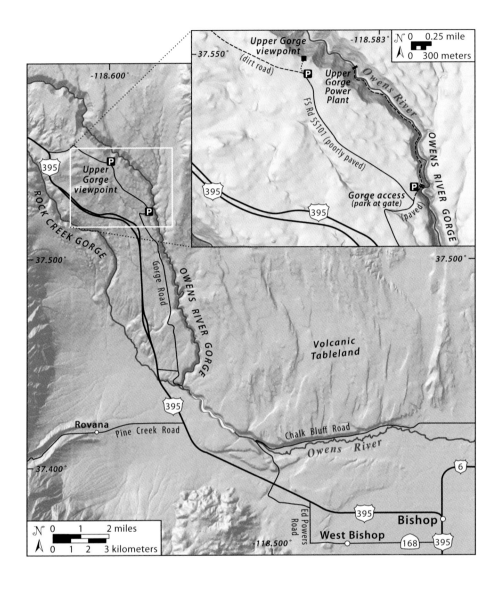

GETTING THERE

Our first of two stops in Owens River Gorge is along the road that goes to the Upper Gorge Power Plant. From the south, drive north along US 395 through Bishop. At the north end of town, US 395 makes a 90-degree westward (left-hand) bend; from this bend, continue 11.8 miles to Gorge Road, which is 2.0 miles past the intersection of US 395 and Mill Creek/Pine Creek Roads. Turn right on Gorge Road and drive 0.7 mile to a T-intersection and turn left. Continue up Gorge Road 6.1 miles to a Y intersection. Take the right fork 0.2 mile to a parking area before a locked gate (37.5253, -118.5752). Park with care; you are next to a 400-foot unguarded cliff. Do not block the gate. The road goes to the Upper Gorge Power Plant (Powerhouse No. 1) and is used regularly by vehicles of the City of Los Angeles Department of Water and Power.

VIGNETTE 27

OWENS RIVER GORGE
A Chasm in the Bishop Tuff

US 395 in eastern California passes through some of the most sublime scenery in the country and provides a stark reminder, unnoticed by most of those who drive over it, of the catastrophic events that shaped the landscape. The finest stretch of the highway skirts the eastern foot of the Sierra Nevada between Fossil Falls (Vignette 18) and Bridgeport. North of Fossil Falls the Sierran crest rises abruptly, reaching its apex at the 14,505-foot summit of Mt. Whitney west of Lone Pine. In the shadow of the high Sierra, the highway follows Owens Valley, climbing gently from about 2,500 feet in the south to about 4,100 feet at Bishop. Water of the Owens River that hasn't been diverted into the Los Angeles Aqueduct flows along the east side of the valley.

North of Bishop, the gentle gradient of Owens Valley is interrupted by the Volcanic Tableland, a large plateau that rises north of town. US 395 must climb nearly 3,000 feet in 10 miles up Sherwin Grade to reach the top of the plateau. The base of the grade lies in desert country of sage and rabbitbrush; the top is in forests of pinyon pine, Jeffrey pine, and juniper. The Owens River emerges from a deep gorge cut into this plateau on Bishop's northern city limit.

The tableland records a geologically recent volcanic eruption of unimaginable violence. About 767,000 years ago, a titanic eruption blasted tephra from a huge elliptical vent area north of Bishop, leaving an enormous depression, the Long Valley caldera. Consolidated debris from this eruption forms a widespread rock unit known as the Bishop

To get to the upper gorge viewpoint, our second stop, return 0.2 mile to the Y-intersection and make a sharp right turn to the north. The road ahead is paved and doesn't get rough until it reaches a "Y" at 1.8 miles. Take the left branch to a large concrete tank where the road splits. Take the left branch, follow it around the left side of the tank to a turnaround circle 0.1 mile beyond the tank (37.5471, -118.6021), and walk east to the rim of the gorge. To return, you may retrace your route, but those with high-clearance vehicles may drive west 2.5 miles to reach US 395.

Tuff. Here we discuss the Bishop Tuff on the Volcanic Tableland, but you should visit Mammoth Mountain (Vignette 30) for a mountaintop view and explanation of the caldera.

The eruption obliterated the preexisting landscape, so we can only infer the events leading up to it. We lack modern analogues because eruptions of this magnitude have not happened since civilization began. Unrest began a few million years ago in the region with volcanic eruptions in what is now Glass Mountain east of Mammoth Lakes. This huge complex of rhyolite domes erupted copious amounts of white tephra. Air-fall tuff layers from these eruptions are widely recognized across southern and eastern California and provide useful time markers, including tuffs in the Tecopa lakebeds (Vignette 19). Presumably the caldera area was a complex of volcanoes, but these are now obliterated and buried.

Before the big eruption, say 800,000 years ago, an ancestral Owens Valley lay between the Sierra Nevada and White Mountains, just as today. Glass Mountain and other pre-caldera volcanoes lay to the north, and before the great eruption the elevation rose 3,000 feet from the level of Owens Valley to the level of Mono Lake, as it does today. An ancestral Owens River flowed south down the valley and, aside from the occasional blanket of white volcanic ash, life was good for the fauna living there—saber-toothed cats, mammoths, bone-crushing dogs, bears, hyenas, camels—the usual Pleistocene gang.

Around 767,000 years ago, however, that good life came to an end. Increasing volcanic activity culminated in a colossal eruption that began by blowing 25 cubic miles of rhyolite tephra into the atmosphere, where it fell like hot snow several feet deep over the area now occupied by the Volcanic Tableland. Recognizable ash from this eruption has been found 1,000 miles away.

This eruption, comparable to the largest in recorded history, was just the first course. It was followed immediately by eruption of about 125 cubic miles of foaming, red-glowing pumice flows that erupted from a series of vents around an elliptical fracture now located within the caldera to the north. Eruption was accompanied by collapse of the area above the emptying magma source, which dropped as much as 2 miles during the eruption. Continued eruption and mega-landslides of the steep walls kept refilling this cauldron, and pyroclastic flows, formed both by fallback of material shot high into the atmosphere and by pumice flows boiling over the rim of the deepening caldera, blanketed the area.

Pyroclastic flows emitted their own billowing clouds of tephra, but most of the frothing, incandescent pumice blocks hugged the ground, racing outward at freeway speeds and grinding each other into sharp glass fragments. About half of this ejected rock material fell back into the caldera, but the remainder blanketed the surrounding area with an outflow sheet of several lobes, all confined between the mountains on either side.

These pyroclastic flow deposits, which erupted at a temperature of about 1,300 degrees Fahrenheit, buried, charred, and boiled plants and animals, and brought a quick end to pastoral life in pre-caldera Owens Valley. Surface water such as that in the ancestral Owens River flashed to steam and rose through the compacting, degassing mixture of pumice and glass fragments, collecting into pipe-like pathways that produced hundreds or thousands of roaring steam vents (fumaroles) at the surface.

The frothy pumice came to rest while it was still hot, pliable, and full of gas bubbles. The tephra accumulated to depths of several hundred feet or more in the area of the tableland and was initially about half gas bubbles. This spongy material compacted under its own weight, squeezing out the gas bubbles, flattening pumice lumps into disks, and producing welded tuff. The frothiest pumice had a density similar to or less than that of water; densely welded tuff is about 2.5 times denser than water, so deeper parts of the pyroclastic flows collapsed to less than half their original thicknesses. This entire pile of compressed pumice is the Bishop Tuff.

According to current interpretations, the main events of the climactic eruption probably took place over about a week. After the pyroclastic flows ceased and the tuff had collapsed and welded, its upper portion consisted of loose ash and pumice, unwelded because there was nothing on top to compress it. Erosion by wind and water quickly stripped this soft and light material away, cutting the surface down to more resistant zones that had undergone some welding. The present surface of the Volcanic Tableland is probably 100 to 200 feet below the original surface.

The total volume of magma erupted in the Bishop Tuff event was about 150 cubic miles. Spread that over the state of California and it would form a frosting 5 feet deep. No volcanic eruption anywhere on planet Earth in all recorded human history has even remotely approached that size, and contemplating the consequences of one to civilization today is sobering indeed. Its volume was some 600 times greater than that of the 1980 eruption of Mt. St. Helens in Washington state, and 50 or more times larger than the eruption of Krakatoa in 1883. Such colossal eruptions were common 20 to 40 million years ago, when dozens of huge tuff deposits blanketed parts of Utah, Nevada, and Colorado. Others built much of the country around and west of Yellowstone National Park over the past 15 million years.

One good way to appreciate such enormous volcanic eruptions is to study the Bishop Tuff on the Volcanic Tableland, where passable paved and graded roads lead to spectacular exposures in Owens River Gorge.

Aerial view looking northeastward over the Owens River Gorge, here about 300 feet deep. Several fumarolic mounds are visible in the foreground. The darker diagonal line just below the center is the Los Angeles Aqueduct, and Gorge Road comes in on the left side of the photo. White Mountain Peak is in the upper right corner.

Much of our understanding of the way in which large pyroclastic flow eruptions and calderas work comes from studies of the Bishop Tuff and the Long Valley caldera by US Geological Survey scientists Wes Hildreth, Roy Bailey, Judy Fierstein, and others, along with New Zealander Colin Wilson. The information contained in the narrative above and the information below comes from them.

Gorge Road turns north at a T-intersection 0.7 mile east of US 395 and follows the large, aboveground, 8-foot-diameter pipe of the Los Angeles Aqueduct for several miles. This pipeline carries most of the Owens River, which falls half a mile in elevation from Lake Crowley to Bishop. The pipeline supplies water to three hydroelectric power-generating stations with a total power output of about 100 megawatts, enough power for about 50,000 typical US homes.

About 1 mile from the T, the road passes by a number of mounds that are typically several tens of feet high. They are the stubs of fumaroles where volcanic steam vented to the surface. Hot mineralized fluids in fumaroles deposit minerals in the surrounding tuff, cementing it and making it more resistant to erosion than uncemented tuff.

If you drive Gorge Road on a clear morning, be sure to stop and admire the magnificent view of the steep mountain range front to your west. A little over 1 mile past the T the road passes just east of one of the taller mounds, and its top is a good vantage point to see huge Mt. Tom, the entrance to the deep canyon of Pine Creek flanked by moraines deposited by a glacier that once flowed out of the canyon, and the steep fault scarp of Wheeler Crest.

The parking area at the Upper Gorge Power Plant provides access to the gorge for climbers and geologists. You are likely to see climbers on the cliffs several hundred feet due south of the parking area, and you might be surprised to see how tiny they look on the massive pillars. After gawking at the gorge, examine the tuff in the roadcuts on the northwest (uphill) side of the road. It is white and porous and consists of chunks of gray and light-brown pumice, or volcanic foam, in a white matrix made up of ground-up pumice—essentially tiny bits of sharp glass. If you look at individual pumice chunks here, you can see that they have roughly equal diameters in all directions. Heft a piece of the rock and note that it is lightweight because it is full of bubbles.

Look closely at the pumice, perhaps with the aid of a hand lens, to see tiny crystals scattered through the glass foam. Clear ones are quartz and

Aerial view of the abrupt south edge of the Volcanic Tableland north of Bishop, looking west. The cliff, in spite of its linear appearance, is not a fault scarp. Rather, it was cut by the Owens River, which exits at the south end of Owens River Gorge at the far end of the cliff, swings sharply and meanders eastward toward the camera, and then resumes its southward flow behind the camera as it nears the base of the White Mountains. Dozens of mostly north-trending normal faults cut the tableland; several are visible here (center right) as dark east-facing scarps. The massive snow-capped pyramid on the skyline left of center is Mt. Tom (13,652 feet). The Tungsten Hills (Vignette 26) are in the middle distance at the foot of Mt. Tom.

In the lightly welded Bishop Tuff (top), photographed above the level of the parking area, the pumice is unflattened and porous. Highly welded tuff (bottom), photographed several hundred feet down the road, has flattened black pumice pieces in a dark-pink matrix. Crystals seem more abundant here than up near the gate, but this is because the porosity has been squeezed out.

feldspar, and black specks are biotite, an iron-rich mica. You might also see small dark bits of older rock, usually less than a few inches across, that were caught up in the magma as it erupted.

Now walk down the powerhouse road past the locked gate. All the rocks exposed in the walls of the gorge are Bishop Tuff. The appearance of the tuff changes as you walk down the road to deeper levels. About 50 feet below the gate, the tuff is composed of blobs of pale pumice, ranging from a few tenths of an inch to over an inch in diameter, in a pink matrix. Here it is noticeably less porous than that examined earlier at the level of the parking lot. The rock is denser yet another 100 feet or so farther down the road, and the pumice clots are flattened to dark-pink streaks. Most of the porosity is gone. About 1,000 feet down the road from the gate, a short distance above where the color changes from pink to dark gray, the pumice clots are black and flattened. Beyond the color change, down in the lower half of the gorge, the tuff is uniformly dark gray and the pumice clots are difficult to see. Heft another piece of rock here, and note how much denser it seems. The bubbles have been squeezed out, and the tuff has lost most of its porosity.

These progressive changes reflect the weight of the overlying tuff. The tuff erupted as an effervescing froth of magma that came to rest as a hot, plastic, porous mass. The weight of the overlying material, coupled with residual heat, compressed the deeper layers into a dense, welded mass, remelting some of the blobs of pumice, flattening others, and turning them black.

CLIMBING IN OWENS RIVER GORGE

Owens River Gorge is a popular spot for rock climbing. The deeper half of the gorge, where the tuff is strongly welded into dark-gray, dense rock, presents a number of challenges. The flattened pumice lumps, so obvious on the walk down the road, weather differently from the rest of the rock. This differential weathering produces hand- and footholds that range from deep pockets that will take your entire foot to tiny edges a fraction of an inch wide that are usable by talented climbers with sticky rubber shoes and strong fingers. This mode of ascent is called face climbing and is the big attraction at the gorge. Most routes have bolted anchors for rope placement.

A climber bypasses a roof on the difficult route Walk the Yellow Line by using tiny ledges along flattened pumice disks that provide purchase on the otherwise smooth face of a huge columnar-jointed block. Rope anchors drilled into the tuff are reused.

Look across to the east wall of the gorge at a world-class example of columnar jointing. Vertical fractures there divide the rock into columns, making it look like a bundle of pencils. Joints are simply fractures; columnar joints like these form when a rock mass contracts as it crystallizes. The fractures are generally oriented so that the columns are at right angles to the surface of the flow and parallel to the direction of heat flow. Devils Postpile (Vignette 31) is another classic example of columnar jointing, and that vignette discusses how they form in more detail.

Spectacular tulip-shaped jointing in poorly welded to nonwelded Bishop Tuff on the east wall of the gorge along the power plant road. Columns are 3 to 5 feet wide. Only about half of the 400-foot-high east wall of the gorge is in this view.

Most of the columns in the east wall of the gorge are remarkably regular and large, typically six-sided and 3 to 10 feet across. At places they curve and converge, like the petals of great stone tulips fanning from their stems, then reaching straight up to the surface. These columnar "flowers" probably formed around fumaroles when the tuff was still hot. Fumaroles are channels that carry hot volcanic gases to the surface. Heat flows outward from them and then up to the surface, just as the radiating columns do. In the lower half of the cliff, in the densely welded, dark-gray zone, the columns are as much as 30 to 60 feet across. They are too crudely formed to be impressive but provide outstanding rock climbing.

Clearly, a river eroded the current gorge, but hardly any water flows through it now. How was it carved? This is best studied at the northern gorge stop a few miles upriver. This complicated story was worked out by Wes Hildreth and Judy Fierstein of the US Geological Survey in a beautiful piece of geologic detective work. They found that the modern

Owens River more or less follows an ancestral Owens River that was obliterated by formation of the caldera. A brief summary of their findings follows.

- An ancestral Owens River carved its course several million years ago in much the same place as the one you are studying. The resulting gorge cut through a ridge of Triassic granite, roughly 220 million years old, and was at least as deep as the present gorge here.
- A small basalt volcano erupted 3.3 million years ago, filled the gorge with more than 600 feet of lava, and spread lava over the local area.
- The lava dammed and diverted the river to the west, where it carved another deep slot in the granite, forming Rock Creek Gorge.
- A few million years later, about 900,000 years ago, glaciers spilling out of the mountains deposited glacial till (coarse debris quarried and transported by glaciers) over the landscape. This till lies on basalt here, showing that the original river course in this area had not been reexcavated by that time.
- Eruption of the Bishop Tuff 767,000 years ago formed Long Valley caldera, and the landscape was blanketed with thick deposits of welded tuff. The tuff lies upon the till and basalt with no evidence that the ancestral river course along this stretch was being reexcavated at that time.
- The caldera floor was at least 2,000 feet lower than the ancestral, basalt-filled river course, which was beheaded at the caldera wall.
- A caldera lake formed. This lake was usually shallow, and it took 600,000 years of intermittent eruption of post-caldera rhyolite flows and domes, along with sedimentation, to fill the basin enough for the lake to overflow. The river from that overflow re-incised the ancient Owens River Gorge to make the canyon before you.

Many of the key actors in this drama are visible from the northern gorge overlook. Directly across the river a ridge comes down toward you. The lowermost rock in the gorge, beneath the basalt and extending to the valley floor, is light-gray Triassic granite. Above the granite are several prominent dark-brown basaltic lava flows that shed dark-brown talus down both sides of the ridge. On top of the basalt is a thin layer of glacial till, inconspicuous but recognizable to the geologist's eye because it sheds large granite boulders down the slope. The resistant top layer with piñon pines upon it is welded Bishop Tuff that lies on light-pink and orange less-welded tuff. This sequence is consistent with the history above: incision of the ancestral river into the granite; filling of that channel with basalt; deposition of till across everything; and deposition of the Bishop Tuff across the till, with no sign of the ancestral river. Things change to the left (north); a series of small faults offset the basalt-granite sequence up to the north, forming a granite ridge that was there before the Bishop Tuff was laid down.

We leave you with one final question. Is it just a coincidence that the present Owens River more or less follows the course of the ancestral, pre-caldera river? Hildreth and Fierstein concluded "probably not." When the caldera formed, the ancestral gorge was likely still a topographic valley north and south of the part that had been filled with basalt. The Bishop Tuff filled this gorge while blanketing the landscape, leaving the Volcanic Tableland as an essentially flat surface, and was several hundred feet thicker at the gorge than elsewhere. A thicker pile compacts and subsides more during welding, thus forming a valley over the former river course, and that valley channeled drainage and focused the course of the reborn river along its former route.

This quiet overlook is an excellent place to eat lunch and ponder the exceedingly violent history of the peaceful landscape before you. Students on a UCLA geology field trip were doing just that in May 1980 when a magnitude 6 earthquake struck with a different kind of geologic violence (Vignette 28).

History of Owens River Gorge revealed at the northern gorge overlook. From bottom to top, Triassic granite, basalt flows (b), till (difficult to see), and Bishop Tuff. The granite and basalt are stepped up to the left along small faults, but the Bishop Tuff is offset less or not at all. The north end of the White Mountains, culminating in white Montgomery Peak, is on the distant skyline at right.

VIGNETTE 28

HILTON CREEK FAULT AT McGEE CANYON
Earthquakes, Boulders, and Gravity

The eastern face of the Sierra Nevada is an imposing escarpment for much of its 400-mile length. That impressive topography was surely accomplished by faulting, but impressive, youthful fault scarps are not evident in the vicinity of the Long Valley caldera with the exception of one spectacular example in the canyon of McGee Creek, just a short distance from US 395 near Lake Crowley. There, the 50-foot-high scarp of the Hilton Creek fault slices through moraines and outwash left by Tahoe- and Tioga-age glaciers. As impressive as this scarp is in McGee Canyon, it cannot be traced very far to the north into the Long Valley caldera. This leaves us with an interesting puzzle. What happens to the fault once it crosses the highway? Does it bend west to follow the mountain face to Mammoth Lakes, or does it just disappear? And if it does disappear, then how? Let's begin a search for the answer first by going to see the fault in McGee Creek.

Turn onto McGee Creek Road from US 395 and head southwest directly toward McGee Mountain. Only a quarter of a mile or so south of US 395, the north end of the Hilton Creek fault scarp is on your right at the base of the planar front of McGee Mountain. The scarp is quite noticeable in the mid- or late afternoon when its face is in shadow.

Upon entering the canyon, the Hilton Creek fault scarp is directly ahead of you as an obvious northeast-facing topographic step behind McGee Creek Campground. The road jogs left to avoid having to climb directly over the scarp, and the campground and small grove of trees use the scarp as a windbreak.

The scarp stands out clearly on the 15,000-year-old northern moraine and in glacial outwash on the valley floor and splits into two parallel strands where it crosses the southern moraine. The fault scarp is about 50 feet high here, and it must be young because it cuts poorly consolidated bouldery glacial deposits and looks barely eroded. Unlike the Lone Pine fault (Vignette 22), slip on the Hilton Creek fault was almost perfectly dip-slip—a classic normal fault with a negligible component of lateral movement.

The scarp formed by accumulation of slip during several earthquakes. A bit of arithmetic: the scarp is about 50 feet high and sharply cuts the moraine crest, which we will assume was last shaped by glacial ice 15,000 years ago. That gives an average slip rate of 0.04 inches per year (1 millimeter per year), which is in line with rates observed on Quaternary normal faults elsewhere in the Basin and Range. How many

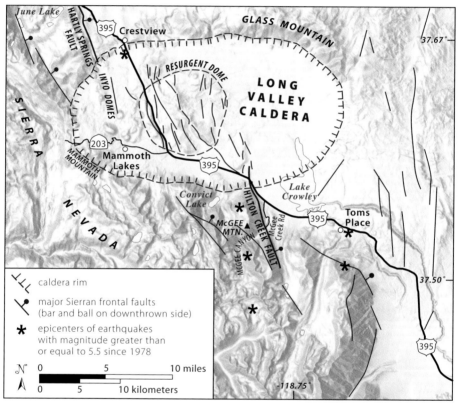

Location of ground breakage (red lines) and approximate epicenters (black stars) in the Long Valley caldera near McGee Creek associated with the four May 1980 earthquakes of magnitude 6. Breakage south of US 395 involved vertical displacement of up to 6 inches along the Hilton Creek fault. The more dispersed breakage within the resurgent dome in the Long Valley caldera north of US 395 was mostly minor horizontal east-west opening of cracks.

GETTING THERE

To reach the Hilton Creek fault scarp, drive to the McGee Creek Road turnoff from US 395 west of Lake Crowley. This road is 6.4 miles northwest of Toms Place and 8.4 miles southeast of the CA 203 turnoff to Mammoth Lakes. Drive 2 miles to the prominent step in the valley floor: the scarp of the Hilton Creek fault.

Aerial view of the Hilton Creek fault scarp cutting across the bottom half of the photo. View is to the southwest, looking upstream into McGee Canyon. The magnificent scarp, nearly 50 feet high, cuts well-developed lateral moraines on each side of McGee Canyon as well as glacial outwash in the valley floor. The broad, flat-bottomed U shape of this valley is typical of canyons carved by glaciers; river-carved canyons have a V shape in cross section. Recreational vehicles in the campground below the scarp give a sense of scale.

earthquakes were involved in producing this scarp? We can't tell from the information at hand, but 10,000 feet of vertical offset could accumulate in 3 million years at that relatively rapid rate.

Recent earthquakes in the area give us information about the nature of this fault system. The Mammoth Lakes region experienced four magnitude 6 (M6) earthquakes over two days in May 1980. Their epicenters were in the mountains south of Convict Lake and west of the Hilton Creek fault. Minor ground breakage occurred in several places along the Hilton Creek fault, and minor cracks were observed farther north in the Long Valley caldera.

Ground breaks and cracks along the Hilton Creek fault ranged from 2 to 6 inches in height where the fault transects the moraines, with the west block on the mountain side of the fault moving up relative to the east block. In some places, vertical separation was up to 10 inches, but some downslope surface slumping may have occurred there, increasing the apparent displacement beyond that caused by fault slip. Farther north, in the caldera north of US 395, breakage was dispersed over a larger area, offsets were small, and displacements produced north-south cracks that opened horizontally rather than vertically.

Did the mountains go up or did the valley go down? The answer is that we cannot tell without an external reference system such as sea level or a geodetic network referenced to the center of the Earth.

This 10-inch-high scarp formed in alluvium at the time of the four M6 earthquakes in the Long Valley area in May 1980. Is it a fault scarp or a slump scarp? It is hard to determine in unconsolidated material like soil and alluvium.

Mountains can go up or down relative to sea level in earthquakes. Geodetic measurements revealed that the San Gabriel Mountains north of Los Angeles rose 8 feet in the 1971 M6.6 Sylmar earthquake, and the Santa Susanna Mountains went up 2 feet in the 1994 M6.7 Northridge earthquake. Conversely, the valley dropped 4 feet along the Lost River fault in the 1983 M6.9 Borah Peak, Idaho, earthquake.

There isn't sufficient data to answer this question for the 1980 earthquakes, but GPS data show that McGee Mountain, on the north side of the creek, rose about an inch between 2006 and 2021, and a point on the range front south of the town of Mammoth Lakes rose about 3 inches in the same period.

The Hilton Creek fault separates alluvium from bedrock along the eastern foot of Mt. Morgan (left), cuts the heads of McGee Creek glacial moraines (center), and cuts across the tops of alluvial fans shed from McGee Mountain (right). Aerial view looking south.

If this ground breakage in 1980 represents the behavior of the Hilton Creek fault over time, then the fault does not turn west around the corner at US 395 and follow the mountain front to Mammoth Lakes as a single fault. Instead, a geologic map published in 1980 by Malcolm Ross and others of the US Geological Survey shows the fault splaying out into the Long Valley caldera as a zone of small, left-stepping faults that transfers displacement westward across the Long Valley caldera to the Hartley Springs fault west of Mono Lake. The change in fault behavior and its orientation at the south edge of the caldera may be a function of the different rock types in the caldera compared to the plutonic and metamorphic rocks of the Sierra Nevada.

Microearthquake swarms, increased fumarolic activity at Hot Creek (Vignette 29), and the surprising coincident release of huge volumes of tree-killing CO_2 at rates equivalent to those at active volcanoes (Vignette 30), suggest that an active magmatic-hydrothermal system deep in the caldera is boiling off aqueous fluids and gas that leak up through fractures in the southwest part of the caldera. As of yet there is no compelling evidence of a cause-effect relationship to the earthquake activity. US Geological Survey seismologists and geologists at the California Volcano Observatory are keeping a close watch on all this activity.

The 1980 earthquakes triggered landslides in and around McGee Creek and demonstrated how earthquakes can reach out and grab you in unexpected ways. Geology students from the University of California, Santa Cruz, were standing near the junction of Convict Lake Road and US 395 when the first quake hit that day in 1980. They watched the rockslides in Convict Creek Canyon as power poles swayed to the Earth's movement. At the same time, Clem Nelson and his UCLA geology field trip class were eating lunch at the northern Owens River Gorge overlook (Vignette 27), with some students sitting on the cliff edge and dangling their feet over the side. Ground shaking there was quite frightening, and students feared being pitched into the gorge. These trips could not have been better timed for some real action!

Boulders crashed into McGee Canyon, barely missing the pack station that lies about 1 mile up the road from the fault scarp. One of the boulders was 10 feet on one side and 3 feet thick; another boulder, about 6 feet across, landed in a corral. Those boulders have since been removed, but several smaller ones remain in and around the pack station, which is still a target for boulders falling from the steep slopes of McGee Mountain. This would be an uncomfortable spot during an earthquake.

The 1997 edition of this book has a view of the fault scarp above McGee Creek Road with a 15-foot boulder embedded high on the scarp

Massive boulder that crashed into the back of site 14 in the McGee Creek Campground. Its presence here is sort of an earthquake double-whammy: it was exposed high in the face of the Hilton Creek fault scarp by repeated earthquakes on that fault, and then dislodged by a M5.3 earthquake in 1998 to end up here.

face. One night in the summer of 1998 an M5.3 earthquake dislodged this boulder, which bounded down the steep slope, punched a deep crater in the road, and rolled into the back of campground site 14. That campsite was occupied at the time, but apparently the boulder was not discovered until morning. It makes a nice climbing target for children and is now just part of the scenery, but it is worth remembering that boulders like this were not placed there by a landscaping service—they fell from the surrounding slopes, and where one has fallen, more are likely to follow.

An especially large rhyolite boulder rolled down the slope behind the old Mammoth School, about 2 miles north of Mammoth Lakes Airport. The boulder is easily visible from US 395, and even better from the first bend in Hot Creek Hatchery Road on the way to Hot Creek (Vignette 29), the former site of the Mammoth School. The boulder came to rest at the foot of the steep slope, which is a fault scarp cut in a rhyolite dome within the caldera. Look for it at the base of the hills to the northeast (to your right as you drive north past the airport turnoff). This fellow measures 25 feet across and left a series of bounce pits on the hillslope, some 6 feet deep, as it bounded down the hill. The pasture around it is littered with large boulders dislodged during earlier earthquakes.

This enormous rhyolite boulder, 25 feet long, bounded down the slope behind the old Mammoth School during one of the 1980 earthquakes, leaving behind a trail of bounce pits still visible forty years later (37.6514, -118.8697). The school was severely damaged by the earthquakes and was subsequently demolished. —Photo by Cecil Patrick

VIGNETTE 29

HOT CREEK GEOLOGICAL SITE
Tempests in a Volcanic Teapot

Hot Creek was once an extremely popular place for bathers to contemplate the phenomenon of magmatic heat while soaking in it. Several vents emitted scalding hot water into the ice-cold creek before 2006, and the place in the creek where bathers congregated was where the cold and hot water blended to about 100 degrees Fahrenheit. Over several decades, the locations of hot vents outside the bathing area moved around, but those in the creek were fairly stable. In May 2006, however, hot springs in and near the bathing area began to spout geysers of blazing hot water several feet above the stream surface at unpredictable intervals. This activity posed a distinct danger to bathers, so the creek was closed to entry in June 2006, and the area was converted into the Hot Creek Geological Site.

Hot Creek's headwaters, under a different name, are in the Mammoth Lakes Basin south of Mammoth Mountain. The creek heads in the Sierra Nevada southeast of Lake Mary as Mammoth Creek, passes through Twin Lakes, and flows eastward between Mammoth Lakes and Old Mammoth into the flat land east of town. This low ground, situated between the topographic margin of the caldera and rhyolite domes that rise in its interior, is the caldera's moat (Vignette 30). The ice-cold trout stream passes under US 395 about 1 mile east of the CA 203 exit, and then a few miles downstream it joins a tributary creek near some hot springs and becomes Hot Creek. It meanders across the meadow for several miles more past a state fish hatchery and the airport, and then enters the 200-foot-deep gorge that hosts the Hot Creek Geological Site. A few more miles downstream, the creek enters flat country once again, this time in the broad eastern moat of the caldera, where it joins the Owens River.

The parking lot is a good place to scout the surrounding landscape. For orientation, the overlook faces northwest. Hills around Hot Creek consist of rhyolite that erupted within the caldera around 330,000 years ago. The White Mountains, capped by lofty White Mountain Peak (14,246 feet), dominate the distant eastern skyline. To the north and northeast, the great mass of Glass Mountain forms the northeastern rim of the caldera. To the west, the top of Mammoth Mountain is visible over

Aerial view looking south and upstream at the Hot Creek Geological Site. The creek meanders across the flat ground in upper right before entering a gorge cut into a 300,000-year-old rhyolite flow.

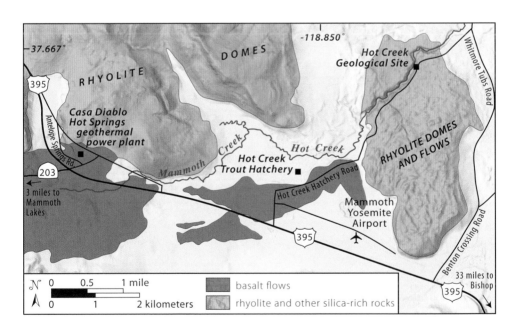

GETTING THERE

Hot Creek Geological Site is located about 3 miles north of the Mammoth-Yosemite Airport. Turn north from US 395 onto Hot Creek Hatchery Road, 3.1 miles east of the intersection of CA 203 and US 395 (Mammoth Lakes exit) and 2.4 miles west of Benton Crossing Road. Follow signs for 3.3 miles to the Hot Creek Geological Site and its paved parking lot.

Hot Creek in 1976. Bathers gathered in the large pool to the right and downstream of the bridge, where icy waters of the stream, coming in from the left, mixed with jets of hot water rising through the creek's sandy bottom. The footbridge was removed in 1997.

Hot Creek in 2021. The former bathing area is the broad section of the stream at left. The blue geothermal pools, part of the ever-shifting plumbing of the geothermal system, formed on the far bank in 2006.

the rhyolite domes. Spectacular summits to the southwest and south surround Convict Lake; the prominent pointed gray monolith almost due south of the parking area is Mt. Morrison (12,270 feet).

As you peer down at the creek from the overlook at the edge of the parking lot, you'll see steam issuing from several fumaroles, most of which are behind protective barriers. The locations and numbers of these vents have changed dramatically over the past several decades. The striking pale-blue pools on the far side of the creek owe their color to various factors. First, they are too hot (hotter than about 160 degrees Fahrenheit) for bacteria and algae to grow. The adjacent darker, greenish pool is cooler and hosts microorganisms. Second, the pale-blue pools are lined with white calcite and clay minerals that reflect sunlight. Finally, finely divided minerals in the water scatter the blue end of the visible spectrum, giving rise to the pretty aqua color.

As you walk down the steep asphalt trail that leads from the parking lot down to the creek, notice the flow-banded volcanic rocks in the walls of the gorge. They are rhyolite lava like that at Obsidian Dome or Panum Crater, but much altered by hot water and volcanic gases. The sulfur-bearing gases, sulfur dioxide and hydrogen sulfide, are particularly corrosive to rocks, vegetation, and geothermal pipes because they combine with water to form sulfuric acid, which attacks volcanic glass and feldspars, turning them to clay. These gases also give many

Flow-banded rhyolite along the Hot Creek trail, once fresh glass like that at Obsidian Dome (Vignette 32), has been chemically attacked by acidic geothermal fluids, and is now a soft, clay-rich mass with remnants of the original flow-banding still present. Camera filter is 2 inches wide.

hot springs their unpleasant, even infernal rotten egg smell, but at Hot Creek the odor is relatively mild. Acid attack has left the rocks around Hot Creek bleached, punky, and converted mostly to clay. An open-pit mine in an inactive geothermal area 2 miles northwest of Hot Creek once produced kaolinite, a white clay used as a filler in products such as paint, plastic, paper, and milk shakes. Not all the rock has been completely altered; in places you may recognize remnants of volcanic glass with attractive flow banding. The rocks at the kaolinite mine provide a preview of what the rocks at Hot Creek will look like one day if they continue to be subjected to corrosive geothermal activity.

Many hot springs surround themselves with deposits of siliceous sinter, a rock made of a form of silica similar to opal, and travertine, a banded form of calcite. Sinter and travertine form when those substances are deposited from hot water as it ascends, cools, and loses its gases. These minerals also deposit in pipes, creating what we might call geo-atherosclerosis and giving headaches to geothermal engineers. Sinter is typically white, but attractive blue varieties are locally common. White sinter and travertine abound along the upper part of the Hot Creek trail at sites of extinct hot springs.

Casa Diablo Hot Springs

Casa Diablo Hot Springs, another geothermal site in the Mammoth area, is about 1,000 feet north of the intersection of US 395 and CA 203. Fumaroles at Casa Diablo Hot Springs served as a spa for centuries, and a long succession of stagecoach stops, trading posts, cafes, and gas stations has occupied this location. Generating stations tap the geothermal power today.

The hillside adjacent to, and north of, Casa Diablo was uplifted along one of the many north- to northwest-trending faults that cut the rhyolite domes in the caldera and channel and localize geothermal fluids and heat. The kaolinite mine northwest of Hot Creek lies along one of these faults. As is quite evident, rhyolitic rocks around here are in a sorry state, altered almost beyond recognition by hot geothermal fluids that still bubble up through the mud. Take care in viewing the many fumaroles that lie near the road, and do not cross any fences.

Geothermal activity at Casa Diablo has been erratic. Generally, the only surface manifestation consists of small, steaming fumaroles similar to those at Hot Creek. On occasion, though, as in 1937 and 1959, geysers spouted hot water and steam as much as 80 feet into the air.

Some geothermal power plants generate steam simply by drilling into hot water, bringing it to the surface, letting it flash to steam at lower pressure in order to drive turbines to make electricity, and then recondensing the steam and pumping it back into the ground. The Casa Diablo plant uses a different technique because the water is not hot enough to do that efficiently. Instead, hot water is brought to the surface and used to boil isobutane, a close relative of butane, the gas used in pocket lighters. The vaporized isobutane drives turbines and is then recondensed to be used again, and the hot water is pumped back into the ground. As of this writing, Casa Diablo has a generating capacity of 30 megawatts, enough to supply around 15,000 homes. A planned expansion will double that. In this way, people are putting volcanic heat to work for them.

Casa Diablo geothermal plant seen looking northwest from the top of a nearby obsidian hill. Fumaroles (F) mark the site of the original hot springs along a north-south fault. The ground (H) along the same fault is barren of vegetation and can be noticeably warm to the touch. Power is generated by turbines (P), and a large condensation plant (C) is used to condense steam to prevent a visible steam plume.

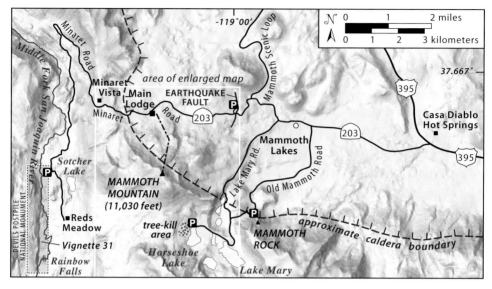
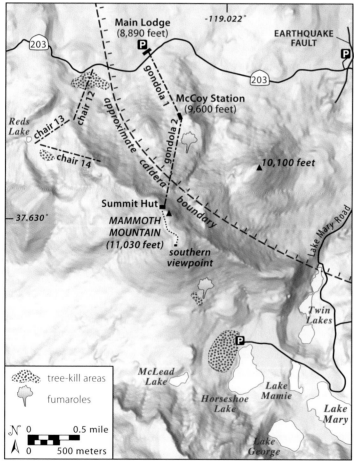

VIGNETTE 30

MAMMOTH MOUNTAIN AND LONG VALLEY CALDERA
Fun and Games on Young Volcanoes

Mammoth Mountain, one of the recreational focal points of eastern California, is a large volcanic dome complex bordering the west flank of the town of Mammoth Lakes. This first-class ski area is heavily used in summer by mountain bikers. Although not particularly lofty by Sierran standards at 11,053 feet, the summit of Mammoth Mountain provides a commanding view of eastern California, the crest of the Sierra Nevada, and the Long Valley caldera and associated volcanoes.

Mammoth Mountain lies along the southwest margin of the Long Valley caldera, a large basin measuring 10 miles north-south by 20 miles east-west. A caldera is a depression left behind when magma rapidly erupts from a subterranean chamber and the unsupported roof of the magma chamber founders. Long Valley caldera is large—you could put San Francisco in there twice—and its size is indicative of a supereruption, an eruption far larger than any that has occurred since the dawn of civilization.

GETTING THERE

This vignette describes three places on and around Mammoth Mountain: the summit; Mammoth Rock, a viewpoint on the caldera rim; and Horseshoe Lake, a place where volcanic gases have produced a ghost forest of dead trees. To reach Mammoth Mountain, drive west on CA 203 (Main Street) 3.7 miles from US 395 through Mammoth Lakes to the Minaret Road turn. This is 1.1 miles past the traffic light at Old Mammoth Road. Turn right and proceed 4 miles to the large parking area by the main ski lodge. Gondola cars run to the summit of Mammoth Mountain most of the year, and the 13-minute ride to the mountaintop is well worth the cost of a ticket. Bring clothes appropriate for windy conditions at more than 11,000 feet because it can be cold even in summer.

Mammoth Rock is near Old Mammoth Road on the south side of town. The trailhead is 3.3 miles down Old Mammoth Road from its intersection with CA 203 on the east side of Mammoth Lakes. Parking is limited (37.6186, -118.9933).

The Horseshoe Lake tree-kill area at the end of Lake Mary Road is 4.9 miles from the CA 203–Minaret Road intersection. Park at the site of a former campground (37.6129, -119.0212), situated among what is now several acres of dead trees.

The brief and intensely violent history of the Long Valley region began about 3 million years ago when magma leaked to the surface, producing a series of stubby domes formed of viscous rhyolitic magma that erupted on and around the present site of the caldera. Many of these eruptions showered eastern California and western Nevada with layers of white volcanic ash, several of which can be recognized across the region in lakebeds such as those at Lake Tecopa (Vignette 19).

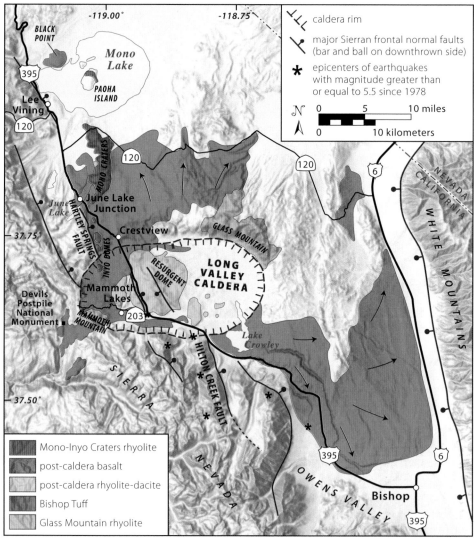

Generalized geologic map of the Long Valley caldera. The caldera margin is especially well defined on its northern, northeastern, and southern margins and is easily seen from the summit of Mammoth Mountain. Basalt lavas form thin flows; rhyolite lava forms prominent domes.

The caldera-forming eruption occurred about 767,000 years ago from a series of ring fractures located around the circumference of the present caldera. Approximately 150 cubic miles of Bishop Tuff blew out in a titanic, week-long eruption. One hundred fifty cubic miles is greater than the volume of Mt. Shasta, and if you piled this tephra on top of an area the size of the present caldera (roughly 150 square miles) it would rise 1 mile high. The caldera roof subsided while the tuff erupted from ring fractures. About half of the material ejected by the blast fell back into the caldera; the other half flowed out in all directions as a series of hot (1,500 degrees Fahrenheit), gas-rich, churning flows of pumice and ash that buried the surrounding landscape. One lobe flowed south down Owens Valley, past the present site of Big Pine; another flowed west, over the Sierran crest and into the San Joaquin River drainage. The remainder blasted high into the air, perhaps as high as 25 miles, where winds carried it eastward and deposited it over most of the western states. Recognizable ash layers are present as far away as eastern Nebraska and Kansas (see map on page 160).

After the eruption pasty rhyolite magma oozed into the caldera, forming domes within it from about 650,000 to 300,000 years ago. Since then, the focus of magmatism has shifted northwestward to the Mono and Inyo Craters (Vignettes 32, 33) and to Mammoth Mountain.

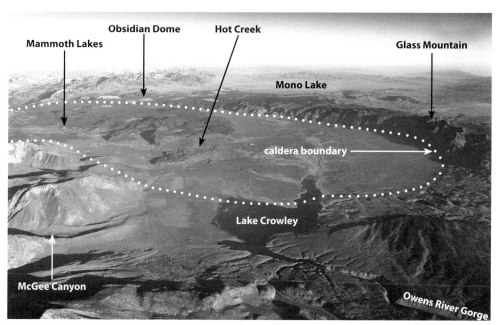

Glass Mountain casts a deep shadow over the northeast margin of the caldera early in the morning in this aerial view looking northwest. Dotted line marks the topographic margin of the caldera. US 395 snakes through the picture from the lower right to Mammoth Lakes and then north past the western shore of Mono Lake.

The gondola to the top of Mammoth Mountain offers stunning views of the caldera and the surrounding terrain even at the start of the ride. To the north, beyond the main lodge, lie unforested rhyolite extrusions of the Inyo and Mono Craters. To the west rises the Ritter Range, easily recognized by the jagged sawtooth ridge known as the Minarets. The caldera lies to the east, and in the far distance are the White Mountains. The view of all these features improves as the gondola car ascends the precipitous northeast side of the mountain.

The ride steepens after passing through McCoy Station as the gondola car ascends the northeast face of the mountain. The steepness of this face is either an expression of a fault that dropped the northeast side of the mountain or perhaps the headwall of a prehistoric landslide. The southwest boundary of the caldera passes through the mountain about here.

You will have a magnificent 360-degree view from the summit. Maps in this vignette plus interpretive photos inside the gondola building will help you identify features. For reference, the Inyo and Mono Craters and Mono Lake are almost due north of the mountain. Consider this

The view from Mammoth Mountain summit looks into Mammoth Lakes Basin. Horseshoe Lake is prominent in the right foreground, and the patch of dead trees on its near side is a major tree-kill area. The lake lacks an overflow outlet as do other lakes in the basin. It drains instead via leakage through its porous bed of volcanic ash, and as a consequence, its level varies dramatically over the seasons. Here in June 2016 the level is low enough to expose a substantial amount of the sandy lakebed. —Photo by Steve Lipshie

THE SIERRA WAVE

Owens Valley and other parts of the east side of the Sierra Nevada are sometimes treated to beautiful cloud formations known locally as the Sierra wave. These long, smooth lenticular clouds follow the Sierran crest a few miles to its east and look something like a stack of white dinner plates or flattened marshmallows. They form when strong west winds blow up the gentle western side of the range and then plunge over the top, down into the adjacent valley, and up the mountains on the other side. This sets up a standing wave pattern, and when the humidity is right, water vapor riding up to the crest of the wave will condense, producing lenticular clouds that are a sure sign of high winds aloft. They can also spawn dangerous rotors—tubular wind patterns in the valley that produce extremely turbulent winds that may be powerful enough to flatten forests on the valley floor.

In the early 1950s the Sierra Wave Project was set up to study the weather and flight characteristics of wave and rotor wind systems. On March 19, 1952, Hal Klieforth and Larry Edgar, flying out of Bishop in a sturdy little Navy surplus glider, caught the Sierra wave during a research flight and set a multiseat glider altitude record of 44,260 feet. Their record stood for a remarkable fifty-four years.

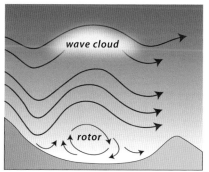

Mechanics of the Sierra wave.

The Sierra wave seen looking south in the vicinity of Fossil Falls at the south end of Owens Valley.

direction to be 12:00. The most obvious and dramatic feature visible from the summit is the Ritter Range, seen at 9:00 to 10:00. Its rugged peaks reach 13,143 feet on sharp-pointed Mt. Ritter. Unlike the more common Sierran peaks, they are carved in Mesozoic volcanic rocks, approximately 140 to 100 million years old, that weather differently from typical Sierran granites.

Moving counterclockwise south (left) of the Minarets, the ground falls away into the drainage of the San Joaquin River, which eventually exits the range near Fresno, about 70 miles southwest. Devils Postpile (Vignette 31) lies hidden along the Middle Fork of the San Joaquin at about 8:30. Farther south, Mammoth Crest and the Mammoth Lakes Basin are visible at about 4:00 to 6:00. Mammoth Lakes Basin was excavated by glaciers where they dug into an older erosion surface, remnants of which you can see on either side of the basin. Mammoth Crest, on the right (west), consists of light-gray granite and looks different from the brown and red rocks of Gold Mountain, which is left (east) of the lakes. The darker rocks are Mesozoic volcanic rocks, more than 100 million years older than the volcanic rocks on which you are standing. Gold Mountain was the focal point of the brief Mammoth gold rush of 1878–80, which attracted as many as 2,500 hopeful gold seekers at its height. The high summits of Mt. Hilgard (13,361 feet) and Mt. Abbot (13,704 feet) peek up on the distant central skyline about 22 miles away.

Left of the lake basin, between about 1:00 and 4:00, the vista opens out into the caldera. The most obvious feature of the caldera margin is the great southwestern face of Glass Mountain, a gigantic pile of obsidian, visible at 2:00 left of and far beyond the town of Mammoth Lakes. Glass Mountain, which forms the northeast boundary of the caldera, consists of lavas that leaked out 1 to 2 million years ago, before the caldera erupted. The lower, forested hills between Glass Mountain and Mammoth Mountain are rhyolite domes that oozed up into the middle of the caldera after it erupted.

Many important features within the caldera are visible, but some are quite subtle. Casa Diablo geothermal area lies at the base of the hills at about 3:00, near the junction of CA 203 and US 395 (the Mammoth Lakes turnoff). Hot Creek geothermal area (Vignette 29), about 5 miles beyond Casa Diablo, is only noticeable if atmospheric conditions allow the fumaroles to make steam plumes. The Huntley Clay Mine is the brilliant white spot in the middle of the rhyolite hills about 10 miles distant. Kaolinite, a white, chalky clay used in paint, paper, ceramics, and milk shakes, was mined there until 2011.

This vantage point gives some clues as to why Mammoth Mountain generally has such deep and dependable snow. It sits alone at the head of a long, low passage up from the Central Valley. The ground falls

View east across the Long Valley caldera from Mammoth Mountain summit. Glass Mountain, the ridge on the center skyline, marks the far margin of the caldera. The White Mountains, more than 40 miles away, form the right skyline.

away dramatically to the west into the valley of the San Joaquin River. The river penetrates deep into the range and passes only a few miles from the Sierran drainage divide near Mammoth Mountain. It is a steep 3,500-foot climb from the river by the Devils Postpile to the summit of the mountain, and 2,000 feet just to clear Mammoth Pass or Minaret Summit, the passes on either side of Mammoth Mountain. Storms heading east over the mountains carry moisture far into the range until they rise and hit this steep final barrier, Mammoth Mountain, and dump their load of moisture as snow—and lots of it.

Stroll south from the summit to obtain an even clearer view of the area to the south and west and to see what the local rocks look like. Most of Mammoth Mountain consists of dacite, pretty purple or gray rocks composed of volcanic glass beset with crystals of glassy quartz, white feldspar, laths of black hornblende, and flakes of black biotite mica. About halfway from the summit hut to the southern view area, however, the dacite is variegated in shades of white, tan, yellow, and orange from alteration by hot fluids and steam. The biotite and hornblende are gone, and the feldspars have turned to a soft, chalky clay. This sort of high-temperature acidic alteration is common around fumaroles (steam vents) and can lead to clay deposits like the Huntley kaolinite pit in the caldera.

For an alternative scenic view of the caldera, take the quarter-mile hike to Mammoth Rock, a tower of white marble that rises 300 feet above the trail. It and the metamorphic rocks along the trail below it were originally marine limestone, siltstone, and sandstone deposited sometime between 350 and 300 million years ago. These rocks are cut by granite dikes and covered in places by a surficial layer of calcite tufa

Mammoth Rock, a 300-foot tower of marble and associated metasedimentary rocks, looms over the Mammoth Rock Trail.

deposited by former springs. The caldera boundary runs through the valley below, and Glass Mountain, 19 miles away on the far side of the caldera, is visible.

The Mammoth Rock Trail is an easy, scenic, and popular path along the south side of the caldera, winding its way to the commercial district along Old Mammoth Road. All the rocks along the trail here and those making up the mountain are old metamorphic rocks and granite, but those in the low ground in front of you are young volcanic rocks, erupted since the caldera-forming event and covered by a veneer of glacial deposits. You must go all the way across the caldera to Glass Mountain, 20 miles away, to find remnants of pre-caldera volcanoes. Rocks that once were adjacent to Mammoth Rock are presumably now 2 miles deep in the caldera right in front of you. You're looking at a landscape that was profoundly changed by eruption of the Bishop Tuff.

After admiring the view from Mammoth Rock, head up to Horseshoe Lake at the south foot of Mammoth Mountain to see the tree-kill area. The trees are gray and dead over an area of sandy shore nearly equal in size to the lake itself. The tree kill there and elsewhere in the region was first noticed in 1989 and was attributed to the drought that gripped California in the late 1980s and early 1990s or to biologic infestation.

Neither explanation was entirely satisfactory, however, because all trees in the kill areas were affected regardless of age, species, or prior health.

The cause of the tree kill was traced to carbon dioxide (CO_2) poisoning, discovered in 1990 when a US Forest Service ranger fell ill with symptoms of suffocation after entering a snow-covered cabin near Horseshoe Lake. Carbon dioxide makes up about 0.03 percent of ordinary air, but measurements taken around Horseshoe Lake after the ranger's illness revealed concentrations more than 1 percent in restrooms and tents and 25 percent in a small cabin. One percent is sufficient to make a person ill, and 25 percent would prove lethal in a short time. Based on the real danger to the public, the Forest Service closed the Horseshoe Lake campground in 1995. It is perfectly safe outdoors, but enclosed indoor spaces are dangerous.

Since 1990, surface emissions of CO_2 following earthquake swarms beneath Mammoth Mountain have at times reached levels as high as those measured around active volcanoes. Persistent ongoing earthquakes as deep as 20 miles indicate that fractures within the mountain are likely active conduits for water vapor and CO_2 making their way to the surface. Hot water escapes at hot springs and fumaroles, and CO_2 escapes invisibly.

Carbon dioxide poisoning killed trees along the north side of Horseshoe Lake. Saplings are growing among the dead trees in the lower left, a sign of recovery. Mammoth Mountain summit is in the background. The blank area in the lower right is the sandy lake bottom exposed in October 2018 by drought.

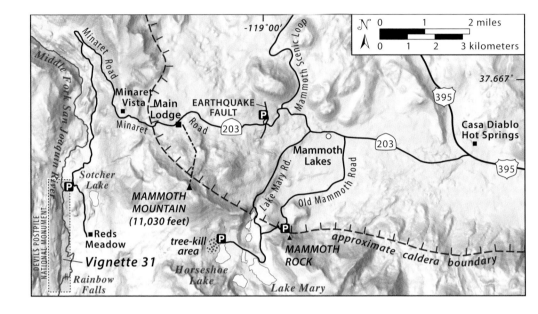

GETTING THERE

To reach Devils Postpile National Monument from US 395, exit west on CA 203, pass through Mammoth Lakes, and head for the ski area. Follow CA 203 past the upper Mammoth Mountain ski lodge and continue to Minaret Summit (9,265 feet), 9.3 miles from US 395. In a normal snow year, the Devils Postpile area is accessible from about mid-June to mid-October. To get there in summer, unless you have a special permit, you will be required to park at the lodge and board a shuttle bus between 7:30 a.m. and 5:30 p.m. daily. For details and bus fares, inquire at the Forest Service visitor center in Mammoth Lakes.

You can reach Minaret Vista, our first stop, by taking the 0.3-mile paved road right (northwest) from Minaret Summit, just before the entrance station. If the shuttle is not operating, you can follow Minaret Road (paved) from Minaret Summit into the valley of the Middle Fork of the San Joaquin River. At a driving distance of 6.6 miles from Minaret Summit, turn right (west) onto a road that leads 0.4 mile to a parking area, visitor center, and the trailhead for a relatively easy 0.4-mile hike to the Devils Postpile. The Middle Fork of the San Joaquin River is a delightful stream, crossed by a footbridge about 0.2 mile upstream from the Devils Postpile. A nice hike of 2.5 miles from the Devils Postpile over easy trails takes you downstream to Rainbow Falls. You may shorten the hike by driving or taking the shuttle bus to the Rainbow Falls trailhead just before Reds Meadow and hiking a 1.25-mile trail to the falls from there.

Collection of specimens (rocks, plants, etc.) within the monument is prohibited. Treat the area gently; you are one of more than 100,000 visitors who come here each year.

VIGNETTE 31

DEVILS POSTPILE
Infernal Patterns in the Rocks

The geometric regularity of columnar jointing is an oddity among the general irregularity of rock outcrops. Well-known localities around the world, such as Fingal's Cave in Scotland, Giant's Causeway in Northern Ireland, Devils Tower in Wyoming, Aldeyjarfoss in Iceland, and Devils Postpile in California are popular and celebrated attractions. All are outstanding examples of columnar jointing in basaltic lava. Basaltic lavas around the globe display a similar remarkably regular pattern of joints, which define parallel polygonal columns a few inches to as much as 20 feet in diameter and up to many tens of feet long. Devils Postpile is a spectacular outdoor laboratory for studying this remarkable process of self-organization.

The Devils Postpile nearly ended up as part of a dam. It was once part of Yosemite National Park, but a boundary change in 1905 left this national treasure on adjacent public land. A proposal to blast the Devils Postpile into the river to make fill for a hydroelectric dam spurred influential California citizens, including John Muir, to persuade the federal government to protect the area. President Taft did so in 1911 by making Devils Postpile a national monument.

On your way to Devils Postpile, take the short side trip to Minaret Vista (see "Getting There"). There you look into the wide, spacious, glaciated valley of the Middle Fork of the San Joaquin River and across it for magnificent views of the jagged Minarets, Mt. Ritter (13,143 feet), and Banner Peak (12,936 feet) of the Ritter Range. These peaks are deformed and steeply tilted, dark-gray volcanic rocks from a Mesozoic caldera. You can see these rocks up close in large boulders at the viewpoint, and they are well-exposed on the drive down to Devils Postpile.

The Middle Fork of the San Joaquin River, which drains the western slope of the Sierra Nevada, is one of just a few major rivers in the Sierra Nevada that flow north-south for part of their course, the Kern River and Bishop Creek being other examples. The Middle Fork of the San Joaquin River may have inherited a north-south orientation in its upper part by eroding a band of old metamorphosed volcanic rocks of similar orientation. The Middle Fork turns sharply west to follow a more normal course about 9 miles downstream from Minaret Vista.

Morning view to the west from Minaret Vista. Right to left, Banner Peak, Mt. Ritter, and the saw-toothed Minarets. All these and the sparsely forested gray outcrops in front of them are steeply tilted, highly deformed volcanic rocks, whereas the more heavily forested light-gray rocks in the foreground are granite.

View of the Devils Postpile from the trail. The scale is a bit hard to appreciate unless there are people visible on top, but the taller columns are 60 to 70 feet high and the tops of the columns about 100 feet above the trail. The massive talus pile in the foreground is composed of broken columns.

After enjoying the view from Minaret Vista, continue to Devils Postpile National Monument, take the 0.4-mile trail to the base of the columns, and behold the grandeur of this strange wonder. You are looking up at thousands of columns packed like a bundle of pencils, some of which are more than 60 feet high. Most are nearly vertical, but they curve to much shallower angles on the left side of the cliff. Some lean out a bit from the wall, and the talus pile below the cliff clearly consists of columns that have fallen and broken into shorter pieces. The rock is basaltic, with abundant small crystals of plagioclase, olivine, and pyroxene.

Columns such as these form when a cooling material contracts and cracks. Imagine that you rapidly fill an Olympic-size swimming pool with basaltic magma and then let it cool. Most of the heat is lost at the top, to the atmosphere; the magma heats the air, which expands, rises, and draws in cooler air, which expands and rises, and so on. This convective process is efficient at drawing heat away from the interior of the lava flow. The pool of lava also cools at the bottom and sides, but heat flow there occurs by conduction and is much slower. As a result, heat flow is largely vertical, out the top of the pool.

The lava pool will rapidly cool down and crystallize, turning the molten material into solid rock with crystals embedded in glass. As the rock continues to cool, it contracts. Basalt cooling from the point at which it solidified to room temperature contracts a few percent. This sets up large tensional stresses in the rock, which responds by cracking.

Forming a new crack in a rock takes a lot of work. That's why rock hammers are big and heavy—to impart a lot of force to the rock. Our

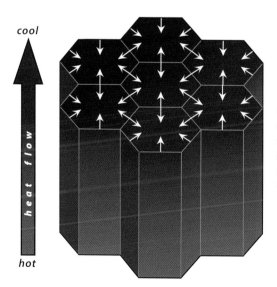

Columns form parallel to the direction of heat flow as the material contracts in a perpendicular direction.

cooling basalt is under a lot of tensional stress from thermal contraction and has to do work (form cracks) to relieve it. Mechanical systems and people both try to accomplish a job by using the minimum work needed, and for our basalt the most efficient way to relieve the stress of thermal contraction is to fracture into hexagonal columns. In other words, the best way to get a crack within a given distance of every point in the rock while forming the smallest total crack surface area is to make hexagonal columns.

The long axes of columns formed this way are parallel to the direction of heat flow. Thus, in our swimming pool most of the columns would be vertical, perpendicular to the surface and bottom, but they would bend toward the sides where heat flowed out the sides. In the Devils Postpile cliff, most of the columns are vertical, but those on the left (north) are shallow to almost horizontal and terminate at a cliff that exposes the bundle of columns end-on. This seems to indicate that this area cooled by horizontal heat flow against something that is not there now. We will return to this puzzle below.

After gazing at the columns, take the trail to the left for the short hike to the top of the Devils Postpile and glaciated bench above and behind

On the north side of the main palisade, the columns curve to near-horizontal orientations and terminate at a vertical surface, which presumably reflects cooling against something that had a vertical surface. What could it have been?

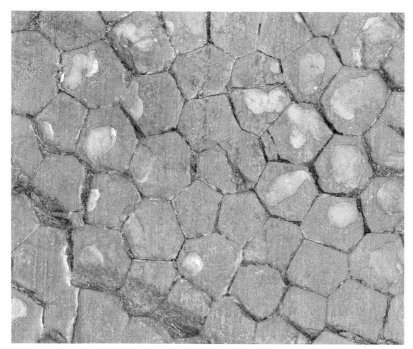

The relatively flat upper surface gives a great view of the geometric precision of these thermal contraction cracks. Note that many columns, which average about 1.5 feet across, are not six-sided. Irregular lighter patches are where the polished and striated surface has weathered away. Like rust under a painted surface, once water gets through the armored coating produced by glacial polishing, it eats outward, eventually destroying the polish.

The upper surface of the Devils Postpile is a great place to contemplate the power of glacial ice. The surface here was ground off and smoothed by ice during the Tioga glaciation. In this view looking south, both the smoothly arched shape of the outcrop and the scratches visible on individual columns were shaped by south-flowing ice.

the cliff, a beautiful, smoothly curving surface that exposes the columns in cross section. Glaciation has clearly shaped the surface, and both the large north-south undulations in the surface and fine north-south scratches were produced by ice flowing down the canyon.

Statistical analyses of the number of sides on columns have been done at columnar jointing sites around the world. A recent count here of nearly 1,000 columns found that about 50 percent have six sides, 25 percent have five, and 25 percent have seven. Thus, hexagons are something that the contracting rock aspires to but does not reach about half the time, and this leads to interesting relationships between the hexagons and non-hexagons. If you have a hexagonal tile pattern and remove one of the hexagons, you cannot fill the hole with a single shape that is not a hexagon. Consequently, the non-hexagons are not randomly distributed, but form chains and clumps. Similar defect chains occur in honeycombs where different teams of bees meet up after starting on different substrates.

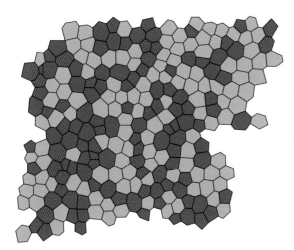

This map of a part of the top surface of the Postpile is colored so that hexagons are gray and 5- and 7-sided polygons (pentagons and heptagons) are red. The latter form chains and clumps that separate areas of hexagons.

The top of the Devils Postpile is about 100 feet above the river, and the basalt continues to the top of the hill a quarter mile to the east, more than 300 feet above the river. Isotopic dating yields an age of around 82,000 years. Thus, at that time the lava erupted from somewhere up-canyon, probably from a vent near Upper Soda Springs campground about 2.5 miles upstream, and the lava seems to have accumulated to a thickness of 350 to 400 feet. That is far thicker than unconfined flows of runny basalt typically attain. Thus, the Devils Postpile flow must have been dammed up by something to become so thick.

What might have dammed the canyon, allowing the lava to accumulate to such great thickness? Given the glacial history of the area, glacial ice or a moraine would be the leading candidates. In particular, lava flowing down a canyon filled with glacial ice will commonly melt its way down the side of the ice sheet, chilling against it. When the glacial ice retreats, the lava flow is left with a steep margin and columnar joints that point at the now-vanished ice.

Damming by glacial ice could explain relationships here, but this hypothesis has problems. The two most recent major ice advances in the Sierra Nevada were the Tioga stage, which culminated about 20,000 years ago, and the Tahoe stage, which culminated about 140,000 years ago. The age of the lava here is in the gap between these major ice advances. There may have been a minor glaciation at this time, evidence for which has been largely erased by the younger Tioga glaciers. Another possibility is that one of the Tahoe glaciers left a moraine in the valley, providing a blockage that choked the valley. The Postpile lava could have banked up against the moraine, which was removed later by Tioga glaciers, leaving behind the much tougher, but also much reduced, lava.

On the south coast of Iceland, about 125 miles east of Reykjavik, near the historic settlement of Kirkjubaejarklaustur (pronounce at your own risk) lies a grassy field. In the midst of the field is a flat area several hundred feet square of glaciated columnar basalt. The pattern there is the same as that in the basalt beneath your feet. It is so attractive that Icelanders say their ancestors built an ancient church at the site, using the glaciated basalt as its floor. The church may have been a mythical structure, but the story attests to the beauty of columnar jointing.

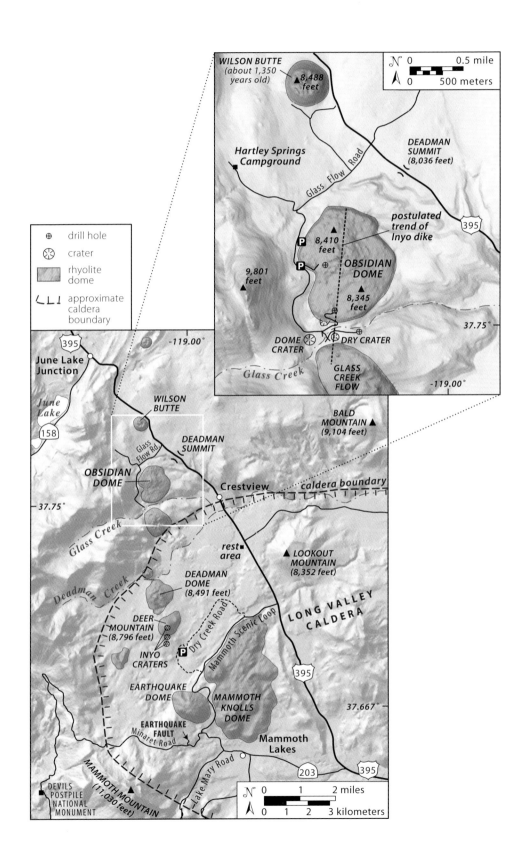

VIGNETTE 32

OBSIDIAN DOME, INYO CRATERS, AND THE EARTHQUAKE FAULT
Medieval Magma Makes Its Mark

It was late summer of the year 1350. The Hundred Years' War between France and England was under way and the Black Death was ravaging Europe; Tenochtitlan, future capital of the Aztec Empire, was newly founded on an island in the lake that would become Mexico City; the Mongol Empire was breaking up; Dante's *Divine Comedy* was on the best seller list; and volcanoes in eastern California, just a few miles north of the future ski town of Mammoth Lakes, were blasting the countryside with ash and spreading pasty lava over the ignited forest. These were not the youngest eruptions in the area—that distinction goes to an event around the year 1700 that injected magma into sediments beneath Mono Lake, lifted them up to produce what is now Paoha Island, and coughed up a small lava flow—but they happened recently enough that the US Geological Survey's California Volcano Observatory categorizes the area as "very high threat potential." In this vignette we examine three sites along the Inyo volcanic chain, a series of domes and related volcanic features stretching north from Mammoth Mountain. From north to south these are Obsidian Dome, a flow of thick rhyolite lava,

GETTING THERE

This vignette examines three features marking recent eruptive activity along the string of volcanic eruptions north of Mammoth Lakes. Together they make a convenient tour of a half day or more.

    To reach Obsidian Dome, proceed north 10.8 miles on US 395 from its junction with the Mammoth Lakes turnoff (CA 203). Turn left (southwest) on Glass Flow Road, 0.3 mile north of Deadman Summit and 4 miles south of June Lake Junction. Drive toward Obsidian Dome, which is visible ahead through the trees left of the road. One mile from US 395, bear left through the intersection with the road to Hartley Springs Campground. You'll turn sharply to the left after the intersection. Four-tenths of a mile farther is a large turnout with a nice view of the dome's margin and various forms of pumice and obsidian. Three-tenths of a mile farther is a parking area that provides access to a road that you may use to walk to the lunar-like top surface of the dome. These roads are drivable with 2WD vehicles, but the pumice-rich surface at the second parking area is locally soft, and vans used for geology field trips have sunk in and gotten stuck.

To reach Inyo Craters, take the Mammoth Scenic Loop (paved) 5 miles north on US 395 from the CA 203 intersection. Turn west and drive 3.1 miles to a dirt road to the right (generally passable with 2WD vehicles) and take it another 1.4 miles to parking with pit toilets. The craters are 0.3 mile from the parking area along a pleasant trail. The turnoff to Inyo Craters is 2.9 miles north of the southern end of Mammoth Scenic Loop at CA 203 (Minaret Road).

To reach the Earthquake Fault, drive 0.9 mile west of Mammoth Scenic Loop on Minaret Road to the signed Earthquake Fault turnoff. It is 1.8 miles west of the intersection of Minaret Road with CA 203. A quarter-mile paved road takes you to a parking area on the north side of the highway, with restrooms and a picnic area. A short hike of a few hundred yards circumnavigates the main trench of the Earthquake Fault.

---

shaped like a mile-wide cow pie, that erupted around 1350; the Inyo Craters, a pair of steam explosion pits formed at the same time; and a crack misleadingly known as the Earthquake Fault that may have been forced open ahead of the rising magma.

Obsidian Dome

Let's begin our tour at Obsidian Dome, an easily accessible example of a young rhyolite flow. Obsidian Dome is not exactly a volcanic dome but rather a steep-sided lava flow known as a coulee (see Vignette 33 for further confusion on the names of volcanic features). Eruption of this lava flow was accompanied by explosive ejection of tephra that winds carried several miles to the northeast, depositing a blanket of air-fall tuff several feet thick near the vent. Fickle winds carried tephra from the next volcano to the south, Glass Creek Flow, to the southwest. Analysis of trees killed by these eruptions, by Connie Millar of the US Forest Service and others, revealed not only the year that the trees were killed, but also the season—late summer, 1350.

Young rhyolite domes are scary to climb upon because their surfaces are literally loose piles of broken glass. Climbing up the side is dangerous, but an old quarry road leads to the top of Obsidian Dome from the second parking area (see "Getting There"), allowing easy access on foot. Obsidian Dome's lava is rhyolite, a volcanic rock with a high concentration of silicon and either few crystals large enough to see with the eye or, in the case of obsidian, few crystals at all. Small crystals form when molten rock cools quickly during an eruption. If this magma had crystallized deep in the Earth, it would have been granite, a coarse-grained rock with the same chemical composition as rhyolite.

VIGNETTE 32: OBSIDIAN DOME, INYO CRATERS, AND THE EARTHQUAKE FAULT

Aerial view looking southwest across US 395 to Obsidian Dome (foreground) and Mammoth Mountain, the snowy peak on the skyline. Other domes of the Inyo chain lie behind and to the left of Obsidian Dome.

Rhyolite magma can take many forms when erupted, and obsidian, pumice, and stony rhyolite are on display here. The obsidian—glossy black glass—contains tiny crystals of feldspar and quartz. The obsidian is mixed with a duller, lighter-colored rock, stony rhyolite, in shades of gray, orange, and pink. Some is pumice, puffed up by bubbles of volcanic gas. Variability comes from differences in how well crystallized the rocks are and in how much gas separated from the magma as it cooled. For an explanation of why obsidian is black, see page 167.

What keeps obsidian from crystallizing? It is a consequence of its high silicon content. In order for a mineral to crystallize from magma, the atoms destined to compose it must migrate through the magma to sites where crystals are forming. In magmas with low concentrations of silicon, such as basalt, this is fairly easy, and basaltic magmas rarely form glass. It is much more difficult for atoms to migrate through a rhyolitic magma because silicon atoms are strongly bonded to one another through oxygen atoms, and these bonds must be broken in order for atoms to move. The more silicon in the magma, the stronger the silicon-oxygen-silicon framework. Rhyolite magma is essentially a thick network of these bonds, very viscous and sluggish. Atoms find it difficult to migrate through this morass, much like swimming through a pool filled with tar. Thus, when magma cools rapidly, the atoms are

Typical black, glassy obsidian layers alternate with frothy gray pumice in this 50-foot cliff on the steep west margin of Obsidian Dome. Beware of sharp edges and falling rocks!

captured in place and do not easily migrate to form crystals. That is why most obsidian has the composition of rhyolite—the higher the silicon content of the magma, the more likely it is to be glassy and the less likely to form crystals.

Volcanoes commonly erupt from point sources, like a broken fire hydrant jetting water; the typical diagrammatic view of a volcano depicts a subterranean magma chamber with a pipe connecting it to the surface. However, magma is not transported in the crust via pipes—it flows through cracks, and when it solidifies in cracks, it forms dikes. The Obsidian, Glass Creek, and Deadman flows erupted in a north-south linear array. Geologists speculated that the various flows vented from a single crack at depth. In 1983 and 1984 they drilled three holes through and near Obsidian Dome to test this idea, the rationale being that knowing how magma is transported through the crust is critical to understanding geothermal resources. One hole was drilled on top of Obsidian Dome, where the road to its top crests, another went through the southern part of the dome, and the third was drilled along Glass Creek. This latter hole, between Obsidian Dome and Glass Creek Flow, was the key.

Banded blocks of obsidian (dark) and pumice (light) are plentiful along the north margin of Obsidian Dome. Fingertip for scale.

It slanted steeply west and hit the feeder dike at a depth of about 2,000 feet, where the dike is 25 feet thick. These observations confirmed the speculation that the chain was fed from a north-trending vertical crack that filled with rhyolite magma.

Inyo Craters

Obsidian Dome and its medieval kin are fine examples of rhyolite lava flows, but the event of 1350 produced other volcanic phenomena as well. At Inyo Craters, steam explosion pits formed when the rising dike encountered groundwater, boiled it, and produced high-pressure steam, which blew to the surface, leaving craters behind—but no lava. The craters are in line with the Inyo chain to the north, and the Earthquake Fault is on the same line to the south. At the Inyo Craters the dike that fed them did not quite reach the surface, but a hole slant-drilled under southern Inyo Crater encountered it 2,000 feet below the crater. Walking around the craters is easy. Scrambling to the top of Deer Mountain is harder but provides an excellent view. Beware of the steep, loose edge around the craters. The cliff on the north side of the south crater has eroded back 10 feet or more in places in the past few decades, leaving the safety fence dangling.

The south and central craters measure more than 600 feet across. The south crater is more than 200 feet deep, and the central about 100 feet deep. The kill age of a log embedded in deposits from the southernmost crater is consistent with the 1350 age for the Obsidian Dome eruption, so these craters are likely part of the same event. To the north at Obsidian Dome and other steep-sided domes, lava poured out of the dike in great volume, but here there is no evidence that magma reached the surface. Instead, we can speculate that early stages of the eruption of Obsidian Dome looked like the Inyo Craters because the rising dike would have encountered groundwater there, too, and blown craters that later filled to overflowing with rhyolite lava.

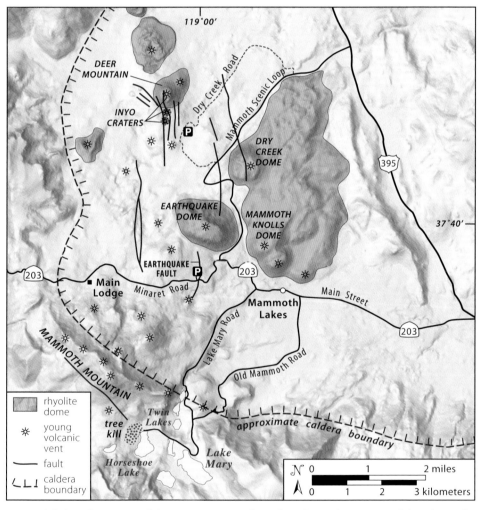

Simplified geologic map of the Inyo Craters and Earthquake Fault. Stars mark locations of young volcanic vents. Earthquake Dome, about 148,000 years old, is far older than the Inyo volcanic chain. The tree kill near Horseshoe Lake is discussed in Vignette 30.

VIGNETTE 32: OBSIDIAN DOME, INYO CRATERS, AND THE EARTHQUAKE FAULT   277

Many north-south normal faults break the area around Inyo Craters, defining narrow, fault-bounded valleys. One of these faults cuts the south rim of the southernmost Inyo Crater, forming an east-facing scarp about 20 feet high. At least some displacement on these faults probably occurred as the ground stretched above the rising Inyo dike. No explosion pits are known south of Inyo Craters, but elongate, undrained depressions and cracks suggest that the dike continues south, toward Mammoth Lakes. One of these cracks is the Earthquake Fault.

Aerial view down and north across the Inyo Craters: south Inyo Crater, which is about 200 feet deep; central Inyo Crater; and Deer Mountain, which has a summit crater of its own. The log that was radiocarbon dated was found embedded in pumice deposits about 20 feet below the rim in the bright cliff above the green lake. The cause of the lake water's typically yellowish-green color is unknown. The central crater contains a lake (out of view) of more normal-looking deep green water. Bare slopes on the left skyline are buried in other rhyolite flows and pumice.

The Earthquake Fault

The Earthquake Fault, an impressively deep and narrow crack in the ground, may be viewed from the bridge across its southern end. The fissure is up to 10 feet wide and 60 to 70 feet deep. We don't know its actual depth because the bottom is filled with loose rock, and ice and snow typically linger year-round in its deeper parts. At one time, steps descended to the bottom, but they were damaged during the Mammoth Lakes earthquakes of 1980 and later removed. Fences now restrict access, but a walk around the main viewing area provides a good look at the fissure.

The Earthquake Fault is not a significant fault in the geologic sense because the rocks on either side, although pulled apart, have not moved very far laterally or vertically relative to one another. *Fissure* would be a more proper term, and although opening of the fissure may have been accompanied by earthquakes, it did not cause them. Take a good look and see if you can visually fit the walls back together. Detailed matching of the walls by students from the University of North Carolina demonstrated that in this area they have separated about 7 feet horizontally, and that the east wall has moved about 3 feet south and 4 feet up relative to the west wall.

The fissure cuts glassy volcanic rocks, part of a rhyolite flow from Mammoth Mountain. Flow banding in the rhyolite, formed by shear when the lava was flowing, inclines very steeply to the west here, and the crack follows it. Although the parallel walls of the crack wiggle back and forth, the overall trend is about due north.

The Earthquake Fault has long attracted attention because large open cracks in the ground are uncommon and may indicate tectonic or landslide activity. A landslide origin can be ruled out because the crack crosses a topographic ridge. That leaves tectonic and magmatic origins as possibilities. The fissure trends north-south and parallels the general pattern of faulting in the area north of Mammoth Mountain. These relationships suggest that the crack is tectonic, related to east-west stretching that is widening the entire Basin and Range province. It also lines up with the Inyo dike, which erupted at Deadman Dome less than 4 miles to the north and reached the water table but did not quite reach the surface at Inyo Craters, 2.4 miles to the north. It's likely that the fissure here formed by cracking ahead of and above the rising dike, which solidified before it reached the water table here. Opening of the fissure ahead of the dike and east-west extension may have happened simultaneously.

The age of the Earthquake Fault is not well known, but its uneroded jagged sides and lack of pumice fill suggest youth. Trees growing in the

VIGNETTE 32: OBSIDIAN DOME, INYO CRATERS, AND THE EARTHQUAKE FAULT   279

The Earthquake Fault in 1981, looking south, shortly before access was restricted by the US Forest Service owing to geologic instability. The floor is 50 feet or more below the rim. The crack is best developed where it follows and splits along steeply dipping flow banding, faintly seen in the shadowed surface on the left edge of the photo. The sunlit surface on the left is such a split; it matches up with the slanted surface below it and to its right. —Photo by Steve Lipshie

Both north and south of the main Earthquake Fault visitor area, the fissure becomes a trough, or series of parallel troughs, before dying out.

fissure indicate a minimum age of about 200 years. With a bit of exploring, you will find that the cleft dies out abruptly both north and south, becoming a mere furrow. It may be that the deep crack just stops, but more likely, it has been filled in by loose rock.

An interesting excursion follows the large fissure south of Minaret Road, where it turns into furrows and shallow fissures. If you search, you may find a mini-Earthquake Fault, inches wide, cutting lava. This trend of furrows and cracks continues south until it is lost under pavement at the west end of Mammoth Lakes.

How threatening is the Earthquake Fault? It lines up with the dike that fed the Inyo volcanic chain and is probably related to that magmatic system. Perhaps one day it will leak molten rock to the surface, but that day, if it comes, could be thousands of years in the future. Perhaps it will never twitch again, slowly filling with debris and fading from sight.

VIGNETTE 33

MONO CRATERS FROM PUNCH BOWL TO PANUM
Mountains of Glass

The chain of twenty-seven or so big domes and three massive flows of glassy rhyolite south of Mono Lake deserve the attention that generations of geologists have lavished on it since the 1889 publication of an elegant monograph on the Mono Basin by an intrepid geologist, Israel C. Russell of the US Geological Survey. We can offer no better introduction to the chain than the one Russell provided in 1889:

> The attention of everyone who enters Mono Valley is at once attracted by the soft, pleasing colors of these craters as well as by the symmetry and beauty of their forms. They are exceptional features in the scenery of the region and are rendered all the more striking by their proximity to the angular peaks and rugged outlines of the High Sierra. The contrast, however, which serves to enhance their beauty tends also to dwarf one's conception of their magnitude. The central cone, crowning the range, rises 2,750 feet above Lake Mono, and has an elevation of 9,480 feet above the level of the sea. The elevation of Vesuvius is about 4,000 feet; the height of Stromboli is a little over 3,000 feet. The larger of the Mono Craters are thus comparable in height with these classic volcanoes of the Mediterranean, but their bases are smaller and their slopes consequently more precipitous. Could this range of craters and lava flows be transported to a region of low relief, as the valley of the Mississippi for instance, it would be far-famed for its magnificent scenery as well as for its geological interest.

Names of volcanic features in eastern California are so mangled that clarification is needed here. The Mono Craters are not, for the most part, craters; they are volcanic domes and coulees. Domes consist of highly viscous, almost solid magma that oozed upward and outward, collapsing as it grew. Coulees are steep-sided, thick lava flows that oozed outward from a vent. Obsidian Dome, Deadman Dome, and Glass Creek Flow (Vignette 32) are also coulees. The Punch Bowl is a crater, and Panum Crater is a dome that mostly fills a crater.

The rhyolitic glass and pumice that compose the Mono Craters volcanic chain began to erupt from fissures at least 35,000 years ago and have continued to erupt episodically up to just 600 years ago. Most of

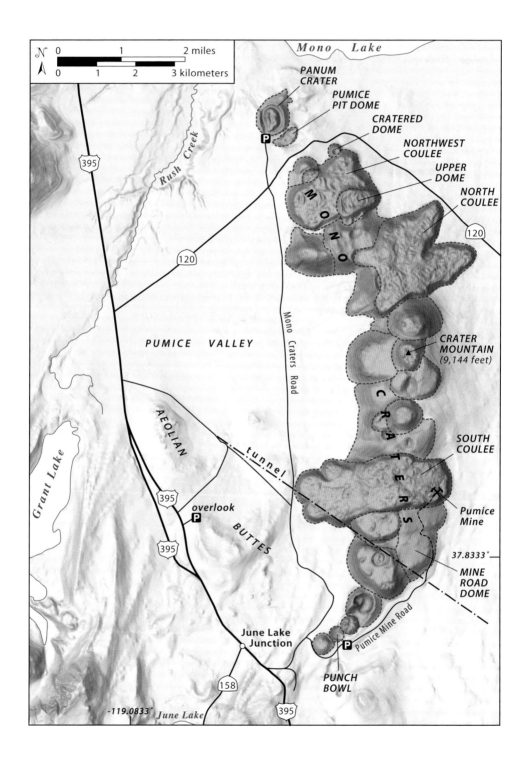

VIGNETTE 33: MONO CRATERS FROM PUNCH BOWL TO PANUM

The Mono Craters are an arcuate chain of rhyolite domes and steep-sided lava flows called coulees. In this aerial view looking south, Panum Crater is in the lower right and CA 120 runs between it and the main Mono chain.

GETTING THERE

This vignette focuses on two volcanic features, one at the north end of the Mono Craters chain and the other at the south. You can approach the domes closely from CA 120, which branches northeastward from US 395 midway along this stretch. The south end of the domes lies just northeast of US 395, 0.5 mile south of June Lake Junction. Hikers in good condition may choose to ascend other larger domes in the chain to their higher parts. Abandoned mining roads and pumice flats provide easy initial access, but climbing steep, pumice-mantled slopes is difficult. Glass-chunk talus slopes and the chaotic surfaces of flows and domes are even more challenging. Exercise care when driving secondary dirt roads near the domes—it's easy to get stuck hub-deep in loose pumice. Stay on well-traveled tracks.

Panum Crater is easily accessible from CA 120. Proceed north on US 395 for 5.7 miles north of June Lake Junction to CA 120 and turn right. Drive 3.1 miles and then turn left (north) at the sign for Panum onto a well-traveled dirt road that leads 0.9 mile to a parking area at the foot of the explosion ring enclosing the crater. Make the short climb up the wide, well-beaten trail to the broad, low saddle in the ring.

To reach the Punch Bowl (37.8149, -119.0259), turn east from US 395 onto Pumice Mine Road, which is 0.5 mile south of June Lake Junction. Follow this wide dirt road and take the left fork in 0.3 mile and then the right fork in another 0.6 mile. The road circles counterclockwise around a rhyolite dome and in 0.9 mile comes to a broad parking area next to a low spot between this dome and the next to the northeast. Park for access to Punch Bowl, which lies in this low spot.

the domes you see are younger than 10,000 years, and many are less than 2,000 years old. The central cluster of highest domes, topping in Crater Mountain at 9,144 feet—2,400 feet above flat Pumice Valley to the west—are among the oldest features in view. The youngest domes and coulees, 600 to 700 years old, cluster at opposite ends of the chain.

The latest volcanic activity recognized so far in Mono Basin is associated with Paoha Island, the large light-colored island in Mono Lake. A small lava flow erupted there about 400 years ago, and a shallow intrusion arched up sediments from the lake bottom, creating the island. In *Roughing It*, Mark Twain tells of how he used the volcanic nature of the area to advantage, boiling gull eggs in hot springs and washing his clothes by dipping them in the alkaline lake waters. For more on Mono Lake and pluvial Lake Russell, the iceberg-laden body of water that filled the basin to a level about 1,000 feet above today's, see *Geology Underfoot in Yosemite National Park*.

Vignette 32 describes a large rhyolite coulee (Obsidian Dome) and two related features, the Inyo Craters steam explosion pits and the Earthquake Fault fissure, which represent embryonic stages in the eruptive history of such a flow. The Mono Craters complete this evolutionary sequence with two features that fill the gap between steam explosion pits, which form when rising magma first nears the surface, and lava flows that form when it has spilled out over the surface.

The Punch Bowl is an explosion crater similar to the Inyo Craters, with two prominent differences: a small rhyolite dome, rather than a lake, occupies the crater's depths, and it has a low rim built of pumice blown from the vent. The crater, about 1,500 feet across and 150 to 200 feet deep, is nestled between two small rhyolite domes, each about 1,800 feet in diameter and 400 feet high. The crater rim rises about 50 feet above its surroundings. The crater and two domes, and another crater about 1,800 feet to the northeast, form a line that trends northeastward and were likely fed by a dike of the same orientation. The Punch Bowl represents an embryonic stage of development in the eruptive sequence. Steam but little to no new lava erupted at Inyo Craters; steam explosions at the Punch Bowl were accompanied by new magma in the form of pumice and the small dome.

Pay attention to the rocks that make up the pumice ring. You will see the usual suspects: rhyolite pumice, stony rhyolite, and obsidian, but you should have no trouble finding rounded pebbles and cobbles of granitic and metamorphic rocks from the Sierra Nevada. These rocks, from the vast alluvial apron through which the Mono Craters erupted, were blasted out of the vent along with the pumice.

The Punch Bowl poses a nice, solvable geologic puzzle: In what order did the Punch Bowl crater and the domes on either side of it form? You

The Punch Bowl is a deep pit with a small rhyolite dome in the bottom. It is bounded on the northeast and southwest by other domes, and on the northwest and southeast by steep ramparts of rhyolite tephra. The taller trees on the dome are about 40 feet tall.

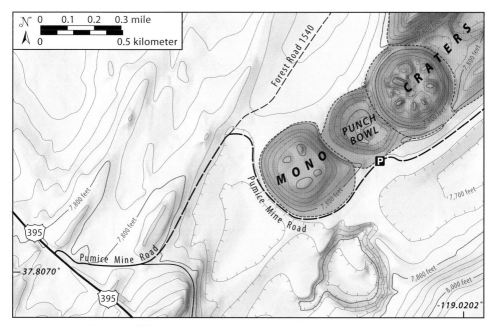

Topographic map of the dome-crater-dome sequence at the Punch Bowl. Tic marks on contour lines indicate depressions. The eruption sequence of the two domes and the crater is obvious from the map once you figure out the key to interpreting it. The contour interval is 40 feet (supplemental contours are in flat area at lower right). —From the US Geological Survey June Lake 7.5′ topographic map

might be tempted to look at tree development to figure this out, but the features probably erupted at most weeks or months apart, so variations in forest cover are likely due to environmental factors such as whether slopes face the sun. The answer is clear from the topographic map once you figure out how to interpret it.

To see the adolescent stage of development of a crater-dome couplet, the next stage beyond the Punch Bowl, proceed to Panum Crater at the north end of the chain. Panum is a favorite stop for tourists and geologists and is being loved to death. Please stay on trails, follow the posted rules, and do not collect anything. Panum's explosion ring is largely intact, and you will recognize the broken-up material, mostly pumice fragments, composing it on your climb up the trail to the saddle in the rim. Rocks in the ring include a scattering of smooth, rounded pebbles and small cobbles of Sierran granitic and metamorphic rocks, as at the Punch Bowl. Panum sits on top of a large gravel delta that sediment-laden Rush Creek built out into a high prehistoric stand of Mono Lake, and the explosions that formed Panum Crater and its ring incorporated some of that deltaic gravel. A splintered glass dome partly fills the center of the Panum bowl. The huge glass domes behind you to the south must have looked something like this in their infancy. You can circle the crater by walking on the crest of the explosion ring, gaining splendid views of Mono Lake and the Sierra Nevada on your stroll.

Although just a whippersnapper, Panum has experienced some upsetting events. After it had cleared its vent of angular rock fragments,

Looking north across Panum Crater to Mono Lake and the Sierra crest. Panum consists of rhyolite, in various shades of gray and orange, and black obsidian. A steep, loose slope of collapsed debris mantles the dome, which is encircled by a sharp-crested rim of pumice. Snow-capped Mt. Dana (13,057 feet) is the highest peak on the skyline.

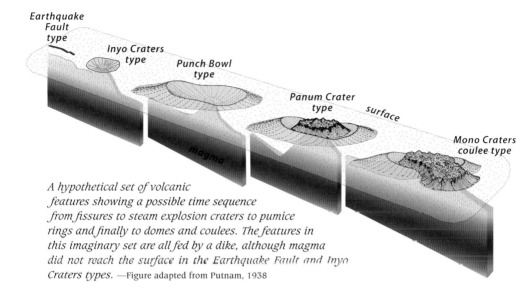

A hypothetical set of volcanic features showing a possible time sequence from fissures to steam explosion craters to pumice rings and finally to domes and coulees. The features in this imaginary set are all fed by a dike, although magma did not reach the surface in the Earthquake Fault and Inyo Craters types. —Figure adapted from Putnam, 1938

erupted sheets of air-fall and outflow pumice, and created the crater, Panum set about building a central glass dome by extruding thick rhyolitic glass. During this process, a second explosion shattered the embryonic dome, blasting a large gap northwestward through the single large explosion ring that encircles the dome, and launching a huge avalanche of big glass blocks into an arm of Mono Lake that occupied the lower part of Rush Creek valley at that time. Succeeding pumice eruptions partly sealed the breached explosion ring, although a gap remains in the highest point of the ring. The wide saddle in the ring upon which you climbed may record a similar but earlier event, even though no trace of ejected material exists outboard of that gap.

Geologists have indirectly dated Panum Crater's dome by the radiocarbon method, using organic sediments interlayered with Panum ash under meadows in the Sierra Nevada. The procedure yields a date for the Panum eruption of about the year AD 1320. This date, subject to the imprecision of radiocarbon dating, is only about thirty years before the Inyo Craters eruptions (Vignette 32) to the south.

A large part of the Mono chain consists of coulees, rather spectacular steep-sided flows that are well worth seeking out. The largest individual effusions of glass within the chain are in the South and North Coulees and a smaller, but still sizable, Northwest Coulee. These rhyolite flows can be as much as 1,000 feet thick. You may gain an excellent view of South Coulee, several domes, and two islands in Mono Lake from a Mono Lake viewpoint reached by traveling 2.1 miles north of June Lake Junction on US 395, then turning right onto a paved road that leads to a

Aerial view of coulees, thick, steep-sided rhyolite flows, spilling down the east side of the Mono Craters. View is to the northwest toward Lee Vining Canyon and the east entrance of Yosemite National Park.

parking area with a kiosk and information signs (37.8423, -119.0681). To return southbound, turn right onto US 395, then left onto West Portal Road. Southbound travelers will have to make a U-turn at June Lake Junction to reach this viewpoint.

You can see the western edge of Northwest Coulee from CA 120 opposite the turnoff to Panum Crater. The best views of the north and south prongs of North Coulee are from CA 120 after it crosses the north end of the chain beyond Panum Crater and curves southeast. The north prong of North Coulee comes close to CA 120, 3.3 miles east of the Panum Crater turnoff, and you may view both prongs from the wide pumice flat one-half mile farther southeast on the highway.

Coulees form where pasty lava oozes from vents and fissures in such great masses that it overwhelms its source and flows outward, submerging the surrounding terrain. They are unusually thick (commonly 200 to 300 feet thick) and steep sided, with talus accumulations at their bases. They flow very slowly, developing a thin, brittle crust that breaks up while still in motion, so that they end up as chaotic jumbles of angular blocks of glass. Sharp, steep-sided glass spires rising above the jumble probably represent shattered remnants of pasty lava squeezed up through cracks in the coulee's broken crust.

Quarry excavations and exploratory drill holes have found permanent masses of solid ice in crevices within domes and coulees to depths of more than 100 feet. This is not surprising because the domes lie at altitudes of up to 9,000 feet, where snow covers them for much of the year. Lots of open space among tumbled blocks, low winter temperatures, and the good insulation provided by pumice and pumiceous rock help preserve the ice. This buried ice in turn cools a relatively low-altitude habitat in a forage-poor environment for colonies of American pikas (*Ochotona princeps*), small rodent-like creatures adapted to the harsh, cold conditions above tree line in the high mountains.

A pika contemplates the various forms that rhyolite can take, its utility for storing hay piles, and the cooling advantages of buried ice from Pleistocene time at Mono Craters. —Photo courtesy of Ken Hickman and Connie Millar, 2020

GLOSSARY

aa. A Hawaiian word for a type of lava with a blocky, jagged, clinkery surface.

air-fall tuff. Volcanic ash deposited by a shower-like falling of pyroclastic fragments from an eruption cloud; drapes the landscape like snow.

alluvial apron. A smooth deposit of alluvium, regional in extent, sloping gently outward from the base of a mountain face. *See also* bajada.

alluvial fan. A low, fan-shaped, gently sloping deposit of alluvium bordering the base of a steep slope at the mouth of a canyon.

alluvium. Unsorted, unconsolidated boulders, cobbles, gravel, sand, and finer rock debris deposited in relatively recent geologic time, principally by running water. Adjective: alluvial.

andesite. A dark-colored, fine-grained volcanic rock intermediate in composition between rhyolite and basalt.

anticline. A fold in layered rocks, convex upward, with stratigraphically older rocks in its core.

ash. *See* volcanic ash.

ash flow. A hot mixture of volcanic gases and ash that travels as a density current down the flanks of a volcano or along the surface of the ground.

ash-flow tuff. A tuff deposited from an ash flow.

axis. As used herein, the center line of a fold in rocks.

badlands. Barren, steep, intricately dissected topography developed by erosion in fine-grained but coherent sediments.

bajada. A broad, continuous alluvial slope extending from the base of mountain ranges out into and around an inland basin, formed by the lateral coalescence of a series of separate alluvial fans.

basalt. A fine-grained, dark, extrusive igneous rock, relatively rich in calcium, iron, and magnesium, and relatively poor in silicon.

basement. The undifferentiated complex of rocks that underlies the rocks of interest in an area.

base surge. A doughnut-shaped cloud of gas and incandescent rock particles that moves outward from the bottom of an explosion column created by a volcanic eruption.

basin. A broad enclosed depression, commonly without drainage to the outside.

Basin and Range. A physiographic province in the western United States characterized by long, linear, fault-block mountains separated by intervening valleys.

batholith. A mass of coarse-grained granitic igneous rock, initially intruded at depths of 2 to 12 miles in the continental crust, and now exposed over a broad area and consisting of two or more plutons.

beach ridge. A low, continuous mound of beach and dune material (sand, gravel, pebbles, shells) heaped up by wave action and currents on the backshore of a beach beyond the present limit of storm waves and the reach of ordinary tides.

bedding. The layered arrangement and structure of sedimentary rocks.

bedrock. The relatively solid rock underlying a mantle of soil or loose rock detritus.

biotite. A common rock-forming mineral of the mica group, usually black or brown, flexible, and with superb platy cleavage.

bluff. A high bank, cliff, or headland rising above a flat.

bomb. *See* volcanic bomb.

boulder. A rock fragment larger than 10 inches in diameter, usually abraded during transport and at least partly rounded.

breccia. A rock consisting of cemented, angular, broken rock fragments.

calcite. A widespread, abundant, typically white mineral composed of calcium carbonate ($CaCO_3$); the major component of limestone and marble. Easily scratched with a knife.

caldera. A large, circular or oval basin formed by collapse following a voluminous volcanic eruption.

caliche. A hard, shallow layer of soil or sediment in which the particles have been cemented together by the precipitation of calcium carbonate.

carbonate rock. A rock composed of the minerals calcite or dolomite, both of which contain carbonate (CO_3); typical examples are limestone, dolostone, and marble.

chert. A hard, dense, dull to partly glassy sedimentary rock composed of finely crystalline silica.

cinder. *See* volcanic cinder.

cinder cone. A cone-shaped accumulation of volcanic cinders erupted from a central basaltic or andesitic vent.

clast. An individual rock fragment produced by mechanical weathering and large enough to be visible to the naked eye in sediment or sedimentary rock. Adjective: clastic.

clay. Rock or mineral particles smaller than 0.002 millimeter, or crystals of a clay mineral.

cleavage. The ability of many minerals to split along crystallographic planes.

coarse-grained. Said of sedimentary rocks that have clasts or particles that are relatively large. Said of igneous rocks with relatively large crystals.

cobble. A rock fragment between 3 and 10 inches in diameter, usually rounded by abrasion in the course of transport.

columnar jointing. The division of rock into prismatic columns by polygonal cracks formed during cooling from a molten state.

conglomerate. A sedimentary rock consisting of rounded pebbles, cobbles, or boulders, cemented within a sandy or silty matrix.

contact. The surface between two types or ages of rocks.

core stone. An ellipsoidal or broadly rectangular joint block of granite formed by subsurface weathering and entirely separated from bedrock.

coulee. A thick, short, steep-sided lava flow, usually rhyolitic and glassy.

crater. A steep-sided, circular depression produced by an explosion.

crust. The uppermost layer of the Earth's lithosphere, typically 3 to 30 miles thick. Continental crust consists mainly of granitic rocks and metamorphic rocks; oceanic crust consists of basalt.

dacite. A medium, light-colored, fine-grained volcanic rock with a composition intermediate between andesite and rhyolite; equivalent in composition to the intrusive rock granodiorite.

debris. A surficial accumulation of loose, broken rock and soil fragments.

debris flow. A relatively fast downslope flow of mixed rock debris as a wet mass.

decomposed granite. Rock detritus, formed by the breakup under weathering of coarse-grained igneous rocks.

deformation. Any process by which preexisting rocks are bent, broken, or uplifted.

desert varnish. A thin coating of dark material abnormally rich in iron and manganese that forms on exposed rock surfaces in desert areas after long exposure.

detachment fault. A complex, low-angle, large-scale normal fault with displacement measured in miles; associated with crustal extension (stretching). Rocks above the fault are displaced on a gently inclined, smooth surface across the underlying rock body of the lower plate.

detritus. A collective term for loose, disaggregated fragments of rock worn off or removed by mechanical abrasion.

dike. A tabular igneous intrusive body, discordant with the structure of the surrounding rock.

diorite. A group of plutonic rocks intermediate in composition between gabbro and granodiorite, and the approximate intrusive equivalent to the extrusive rock andesite.

dip. The inclination from horizontal of any planar surface within rocks, such as a sedimentary bed, as measured in the steepest direction (e.g., the direction a marble would roll down an inclined surface).

dip-slip fault. A fault along which the blocks have been displaced vertically relative to each other. Normal and reverse faults are dip-slip faults.

dolomite. A common rock-forming carbonate mineral with the formula $CaMg(CO_3)_2$; term is also applied to a sedimentary rock consisting of dolomite.

dolostone. A sedimentary rock composed primarily of dolomite.

epicenter. The point on the Earth's surface directly above the point of origin of an earthquake.

erosion. The mechanical destruction and removal of rock material by natural processes.

erratic. A rock fragment carried by glacial ice, deposited at some distance from the outcrop from which it was derived, and typically but not necessarily deposited upon a bedrock of different rock type. Size ranges from a pebble to a house-size block.

extrusive. Said of igneous rock that has erupted onto the surface of the Earth, including lava and pyroclastic material such as volcanic ash.

fanglomerate. The consolidated deposits of an alluvial fan; a variety of conglomerate that is coarse, poorly sorted, only weakly bedded, and contains angular stones.

fault. A fracture along which blocks of Earth's crust have slipped past each other.

feldspar. A group of common rock-forming minerals composed principally of silicon, aluminum, and oxygen, plus one or more of the elements calcium, sodium, and potassium.

fine-grained. Said of sedimentary rocks that have clasts or particles that are relatively small. Said of igneous rocks with relatively small crystals.

fold. Bent or warped rock layers shortened by compressive tectonic stress.

formation. Geologically, a rock body of considerable extent with consistent characteristics that permit it to be recognized and mapped; not to be confused with outcrop.

fossil. The remains of a plant or animal preserved in a rock.

fossiliferous. Said of a rock containing fossils.

fracture. Any break in rocks caused by natural mechanical failure under stress, including cracks, joints, and faults.

fumarole. A small volcanic vent that emits hot fluid or gas.

gabbro. A coarse-grained, dark intrusive igneous rock formed from the slow cooling of magnesium-rich and iron-rich magma. It is chemically equivalent to rapid-cooling, fine-grained basalt.

geothermal. Involving heat from within the Earth.

glacier. A large body of natural, land-borne ice that flows.

glaciation. The formation and movement of glaciers or large ice sheets.

gneiss. A regionally metamorphosed rock characterized by alternating bands of coarse mineral grains and finer, flaky mica minerals. Mineral composition is not an essential factor in its definition; varieties are distinguished more by texture and origin.

granite. An igneous intrusive rock consisting mostly of visible crystals of quartz and feldspar.

granodiorite. An intrusive igneous rock midway in composition and color between granite and diorite.

gravel. An unconsolidated accumulation of rounded rock fragments, mostly of particles larger than sand, such as cobbles and pebbles.

groundwater. The water in pores and other openings in subsurface rocks and sediment.

hornblende. A general term for a complex mineral of the amphibole series, rich in calcium, magnesium, and iron. It commonly forms prominent black crystals in granite and granodiorite.

hydrothermal. Pertaining to geothermally heated water.

ice age. A period in Earth history when large sheets of ice covered parts of nonpolar continents.

igneous. Said of rocks and minerals formed by the crystallization of molten material (magma).

intrusive rock. Rock formed by crystallization of magma underground.

isotope. A species of an element defined by the number of neutrons in its nucleus. Adjective: isotopic.

isotopic dating. Determining the age of a geologic sample by measuring isotopic ratios of parent to decay products or the remaining amount of a radioactive isotope.

joint. A planar fracture in a rock without displacement; commonly in parallel sets.

lakebeds. Fine-grained sedimentary deposits laid down on the floor of a lake.

lava. Extruded magma or the solidified product of such.

left-slip fault. A strike-slip fault along which the opposing block moved to the left.

limb. That area of a fold between adjacent fold hinges.

limestone. A sedimentary rock composed largely of the mineral calcite ($CaCO_3$).

lithosphere. The rigid outer part of the Earth; includes the crust and upper mantle.

longitudinal dune. A long, narrow dune ridge parallel to prevailing wind.

magma. Naturally occurring molten rock within the Earth, capable of intrusion and extrusion. Once extruded, it is called lava.

magnitude. For earthquakes, a measure of strain energy released abruptly along a fault. Magnitude is measured on a logarithmic scale, with each increase of one unit of magnitude corresponding to a tenfold increase in the amplitude of ground shaking, and a thirtyfold increase in the energy released.

mantle. The plastic zone of the Earth between the core and the crust.

marble. A metamorphosed limestone or dolomite, usually coarsely recrystallized.

marine. Pertaining to sedimentary rocks formed in the sea or ocean, and to oceanic environmental conditions.

matrix. Fine-grained rock or mineral particles filling spaces between coarser constituents of a sedimentary rock.

megabreccia. A term for a deposit of rock produced by brecciation (cracking) on a large scale and containing randomly oriented blocks that range from a few feet to more than 300 feet in horizontal dimension.

metamorphic rock. A rock that has undergone sufficient mineralogical and physical changes by heat, pressure, and shearing stress to be distinct from the parent rock. The prefix *meta-* is added to other rock names to indicate they've been metamorphosed.

metamorphism. Recrystallization of an existing rock due to heat and pressure within the Earth.

mica. A group of rock-forming silicate minerals with perfect sheetlike cleavage; includes biotite and muscovite.

moat (volcanic). A gully-like depression between a volcanic dome and its explosion rim, or between a resurgent dome and caldera margin.

moraine. An accumulation of poorly sorted rock debris carried downslope by a glacier.

mudstone. A fine-grained sedimentary rock of silt and clay, coarser-grained and more massive than shale; hardened mud.

narrows. Topographically, a gorge or constricted passage along a stream or through a pass.

normal fault. An inclined fault on which the block of rock above the fault has moved relatively downward.

North American Plate. One of eight or more huge moving plates that make up the outer, solid part of the Earth, the lithosphere.

obsidian. A black or dark-colored volcanic glass; usually rhyolite lava that cooled too quickly to form crystals.

olivine. An iron and magnesium silicate mineral that typically forms pale-green to pistachio-green crystals in basalt, gabbro, and peridotite.

outcrop. An exposure of bedrock at the Earth's surface. The rock is said to "crop out."

Pacific Plate. One of Earth's major plates, lying west of the North American Plate and separated from it, in California, by the San Andreas fault.

pahoehoe. A Hawaiian term for a type of basaltic lava typified by a relatively smooth, bulbous, ropy surface.

paleontology. The study of ancient life, largely by means of fossils.

pebble. A small, rounded stone, usually waterworn in the course of transport, between 0.17 and 2.5 inches in diameter.

peridotite. A coarse-grained igneous rock, poor in silicon and rich in iron and magnesium, consisting mainly of olivine. The Earth's mantle consists of peridotite.

petroglyph. A carving on a rock surface, commonly formed by pecking away desert varnish.

piping. The formation of voids within rock or soil caused by the removal of material by seepage.

plagioclase. A feldspar mineral rich in sodium and calcium. One of the most common rock-forming minerals in igneous and metamorphic rocks.

plate. In a planetary sense, large, drifting plates composing the Earth's solid outer part, the lithosphere. Continental plates are roughly 60 miles thick.

plate tectonics. The movement and deformation of the Earth's crust caused by the interaction of planetary plates.

playa. A smooth, fine-grained lakebed in a desert valley, normally dry but may be flooded.

pluton. An intrusion of igneous rock into overlying rock. Named for Pluto, the Greek god of the underworld.

pluvial. Pertaining to cooler, moister conditions in arid or semiarid areas, possibly coincident with glacial conditions in other regions.

polygon. A closed geometric figure with three or more straight sides.

pothole. A cylindrical hole drilled into the rock bed of a high-velocity stream by a fixed vortex armed with sand and gravel.

pumice. A highly bubble-filled volcanic glass, sometimes light enough to float on water.

pyroclastic. Said of broken rock material and pumice formed by explosion from a volcanic vent.

pyroxene. An iron- and magnesium-bearing silicate group of minerals that occur mostly in dark, low-silica igneous and metamorphic rocks.

quartz. A common rock-forming mineral that is hard, chemically resistant, and composed of silicon and oxygen (SiO_2). Cannot be scratched with a knife.

quartzite. A metamorphic rock formed by the recrystallization of quartz-rich sandstone.

radiocarbon dating. Dating using a radioactive isotope of carbon, which disintegrates with a half-life (the time required for half of the isotope to decay) of 5,730 years.

resurgent dome. A collection of domes within a caldera formed as magma welled up after the caldera-forming eruption.

reverse fault. An inclined fault, typically steeply dipping, with relative upward movement of the block of rock above the fault.

rhyolite. An extrusive igneous rock (lava) of granitic composition, commonly light-colored or reddish. Relatively rich in silicon and poor in iron, magnesium, and calcium.

right-slip fault. A strike-slip fault along which the opposing block moved to the right. Also: right-lateral strike-slip fault.

sag pond. A small body of water occupying an enclosed depression or sag formed where active or recent fault displacement has impounded drainage.

salt pan. A large flat area in which salt water accumulates and evaporates, leaving a layer of snow-white salt crystals.

sandstone. A sedimentary rock composed primarily of rounded or angular, sand-size particles of rock or mineral, 0.0025 to 0.08 inch in diameter.

scarp. Topographically, a steep cliff face, from a few to thousands of feet high, commonly produced by faulting.

schist. A metamorphic rock characterized by strong cleavage, usually involving similarly oriented mica flakes.

sediment. Solid, unconsolidated particulate matter, especially rock detritus that originates by weathering and is transported and deposited by wind or water.

sedimentation. The process of deriving, transporting, and depositing sediment.

sedimentary rock. Consolidated cemented sediment, characterized by layering.

shale. A consolidated sedimentary deposit of clay, silt, or mud.

silica. Silicon dioxide (SiO_2).

silicate. A mineral compound whose crystal structure contains SiO_4 tetrahedra, either isolated or joined with metallic elements through one or more of the oxygen atoms to form groups, chains, sheets, or three-dimensional structures.

silt. Fine particulate rock and mineral matter, dust-size (finer than sand, coarser than clay), between 0.00016 and 0.0025 inch in diameter.

siltstone. A sedimentary rock composed primarily of silt.

slip. The amount of displacement along a fault.

sorting. The arrangement of sedimentary rock particles naturally selected by size, shape, or specific gravity by agents of transportation. Adjective: sorted.

star dune. An individual dune with several ridges radiating out from a central high point, resembling a starfish when viewed from above.

steptoe. An isolated hill or mountain of older rock surrounded by a lava flow.

strandline. A former shoreline of a body of standing water.

stratification. Layering in sedimentary rocks.

stratigraphy. Pertaining to the stacking or sequence of deposition of sedimentary rock layers.

strike. The compass direction of a horizontal line with an inclined rock layer.

strike-slip fault. A fault along which the relative displacement is sideways rather than up or down.

syncline. A U-shaped downfold in layered rocks with younger beds toward the core, limbs inclined inward.

Tahoe glaciation. The local name for the penultimate stage of glaciation in the Sierra Nevada, which culminated about 140,000 years ago; older than the Tioga glaciation.

talus. The accumulation of large angular blocks of rock at the base of a cliff.

tectonic. Pertaining to deformation and the forces that cause deformation of the Earth's crust.

tephra. Fragmental products of magma formed by, and ejected during, volcanic eruptions.

thrust fault. A fault with a dip of 45 degrees or less on which the upper block has moved upward relative to the lower block.

Tioga glaciation. The local term for the youngest major phase of glaciation in the Sierra Nevada, which culminated about 20,000 years ago.

trace fossil. A structure, such as a track or burrow, that is left by the activity of a plant or animal.

transverse dune. An asymmetric sand dune elongated perpendicular to the prevailing wind direction.

travertine. A dense accumulation of calcium carbonate resulting from deposition by groundwater or surface water.

tufa. A rock composed mainly of calcium carbonate deposited from water and found at spring sites or along strandlines of saline lakes.

tuff. A volcanic rock formed of consolidated tephra.

tumulus. A dome or small mound in the crust of a lava flow.

turtleback. A geologic structure, best known from Death Valley, featuring a smooth, denuded rock surface with the shape of a plunging anticline, resembling the shell of a turtle.

unconformity. A surface of erosion or nondeposition separating younger deposits from older rock. *See* angular unconformity.

vein. A sheetlike deposit of mineral matter within a fracture in rock.

vent (volcanic). A roughly cylindrical opening through which volcanic material is extruded.

ventifact. A stone whose surface and shape have been modified by windblown sand.

vesicle. A small cavity of irregular to spherical shape formed by a trapped gas bubble in lava. Adjective: vesicular.

viscous. Said of the state of a fluid with a cohesive, sticky consistency. A viscous fluid flows sluggishly; a nonviscous fluid flows easily.

volcanic ash. Unconsolidated, explosively fragmented, pyroclastic volcanic material of particle diameter less than 0.25 inch.

volcanic bomb. A glob of lava ejected while still viscous and shaped and cooled in flight. Shape, form, and size vary greatly.

volcanic cinder. A glassy, porous fragment of lava explosively ejected from a volcanic vent; from pea to baseball size.

water table. The top of the subsurface zone that is saturated with water.

weathering. The chemical decomposition and mechanical disintegration of rocks and minerals through interaction with the atmosphere and biosphere.

welded tuff. A glass-rich, coherent, fragmented volcanic rock, hardened partly by the heat of its constituents.

wind ripple. A wavelike, asymmetric undulation produced in wind-deposited sand.

xenolith. A foreign rock fragment included in an igneous rock.

LIST OF SOURCES

Bedinger, M. S., and J. R. Harrill. 2012. *Groundwater Geology and Hydrology of Death Valley National Park, California and Nevada.* Natural Resource Technical Report NPS/NRSS/WRD/NRTR—2012/652. National Park Service. Available online.

Bryan, T. S., and B. Tucker-Bryan. 1995. *The Explorer's Guide to Death Valley National Park.* Niwot: The University Press of Colorado.

Digonnet, M. 1999. *Hiking Death Valley: Guide to its Natural Wonders and Mining Past.* Palo Alto, CA: Wilderness Press.

Gebhardt, C. 1988. *Inside Death Valley*, 4th Edition. San Francisco, CA: Tom Willis Press.

Lingenfelter, R. E. 1986. *Death Valley and the Amargosa.* Berkeley: University of California Press.

Lipshie, S. R. 2018. *Geologic Guidebook to the Long Valley: Mono Craters Area of Eastern California, 3rd Edition.* Reno: Geological Society of Nevada. Available from gsn@gsnv.org.

Miller, M., and L. A. Wright. 2015. *Geology of Death Valley National Park, 3rd Edition.* Dubuque. IA: Kendall/Hunt Publishing Company.

Sylvester, A. G., and E. O. Gans. 2016. *Roadside Geology of Southern California.* Missoula, MT: Mountain Press Publishing Company.

Sylvester, A. G., R. P. Sharp, and A. F. Glazner. 2020. *Geology Underfoot in Southern California*, 2nd Edition. Missoula, MT: Mountain Press Publishing Company.

Walker, J. D., J. W. Geissman, S. A. Bowring, and L. E. Babcock, compilers. 2018. *Geologic Time Scale* v. 5.0, Geological Society of America, available online.

1. BADWATER

Brogan, G. E., K. S. Kellogg, D. B. Slemmons, and C. L. Terhune. 1991. *Late Quaternary Faulting Along the Death Valley–Furnace Creek Fault System, California and Nevada.* US Geological Survey Bulletin 1991.

Frankel, K. L., L. A. Owen, J. F. Dolan, and others. 2015. Timing and rates of Holocene normal faulting along the Black Mountains fault zone, Death Valley. *Lithosphere* 7 (6).

2. DEVILS GOLF COURSE AND BADWATER SALT FLATS

Lasser, J., J. M. Nield, M. Ernst, and others. 2019. Salt polygons are caused by convection. *arXiv:* 1902.03600v2.

3. NATURAL BRIDGE CANYON

Hall, S. A. 2020. A decade of discovery in Death Valley National Park. *Span* 32 (3): 6–10. Available online.

LeBlanc, K. E., and J. R. Knott. 2010. *Using repeat photography to document landscape change in Death Valley over the last 100 years.* AAPG Search and Discovery Article #90114©2010 AAPG Pacific Section Meeting, Anaheim, California.

Miller, M. G. 1991. High-angle origin of the currently low-angle Badwater turtleback fault, Death Valley, California. *Geology* 19: 372–75.

4. VENTIFACT RIDGE

Roof, S., and C. Callagan. 2003. The climate of Death Valley, California. *Bulletin of the American Meteorological Society* 84 (12): 1725–39.

5. ARTISTS DRIVE

Knott, J. R., and A. M. Sarna-Wojcicki. 2001. Late Pliocene tephrostratigraphy and geomorphic development of the Artists Drive structural block. In *Quaternary and Late Pliocene Geology of the Death Valley Region: Recent Observations on Tectonics, Stratigraphy, and Lake Cycles*, eds. M. N. Machette, M. L. Johnson, and J. L. Slate, US Geological Survey Open-File Report 01-51, p. C105–C116.

6. GOWER GULCH

Crippen, J. R., 1979. *Potential Hazards from Floodflows and Debris Movement in the Furnace Creek area, Death Valley National Monument, California-Nevada*. US Geological Survey Open-File Report 79-991.

Pistrang, M. A., and F. Kunkel. 1964. *A Brief Geologic and Hydrologic Reconnaissance of the Furnace Creek Wash Area, Death Valley National Monument, California*. US Geological Survey Water-Supply Paper 1779-Y.

Snyder, N. P., and L. Kammer, 2008. Dynamic adjustments in channel width in response to a forced diversion: Gower Gulch, Death Valley National Park, California. *Geology* 36 (10): 187–90.

Troxel, B. W. 1973. Significance of a man-made diversion of Furnace Creek Wash at Zabriskie Point, Death Valley, California. *California Geology* 27 (10): 219–23.

7. DANTES VIEW AND TURTLEBACKS

Cowan, D. S., T. T. Cladouhos, and J. K. Morgan. 2003. Structural geology and kinematic history of rocks formed along low-angle normal faults, Death Valley, California. *Geological Society of America Bulletin* 115 (10): 1230–48.

Hamilton, W. B. 1988. Detachment faulting in the Death Valley region, California and Nevada. *US Geological Survey Bulletin* 1790: 51–85.

Miller, M. G. 1991. High-angle origin of the currently low-angle Badwater turtleback fault, Death Valley, California. *Geology* 19: 372–75.

Stewart, J. H. 1967. Possible large right-lateral displacement along fault and shear zones in Death Valley–Las Vegas area, California and Nevada. *Geological Society of America Bulletin* 78 (2): 131–42.

Stewart, J. H., and F. G. Poole. 1974. Lower Paleozoic and uppermost Precambrian Cordilleran miogeocline, Great Basin, western United States. In *Tectonics and Sedimentation*, ed. W. R. Dickinson, SEPM Special Publication 22, p. 28–57.

Wernicke, B. P., J. K. Snow, G. J. Axen, and others. 1989. *Extensional Tectonics in the Basin and Range Province between the Sierra Nevada and the Colorado Plateau*. International Geological Congress Field Trip Guidebook T138, American Geophysical Union, Washington, DC.

8. LAKE MANLY

Blackwelder, E. 1933. Lake Manly: An extinct lake of Death Valley. *Geographical Reviews* 23 (3): 464–471.

Knott, J. R., J. M. Fantozzi, K. M. Ferguson, and others. 2012. Paleowind velocity and paleocurrents of pluvial Lake Manly, Death Valley, USA. *Quaternary Research* 78 (2): 363–72.

9. MESQUITE DUNES

Haff, P. K. 1986. Booming Dunes: For reasons not yet fully understood, sand sliding down the slipface of a dune can produce booming noises at startling volumes. *American Scientist* 74 (1): 376–81.

Smith, H. T. U. 1967. *Past Versus Present Wind Action in the Mojave Desert Region, California.* US Air Force Cambridge Research Labs Publication AFCRL-67 0683.

10. MOSAIC CANYON

Beaumont, P., and T. M. Oberlander. 1971. Observations on stream discharge and competence at Mosaic Canyon, Death Valley, California. *Geological Society of America Bulletin* 82: 1695–98.

Hodges, K. V., J. D. Walker, and B. P. Wernicke. 1987. Footwall structural evolution of the Tucki Mountain detachment system, Death Valley region, southeastern California. In *Continental Extension Tectonics*, M. P. Coward and J. F. Dewey. Geological Society of America Special Publication 28, p. 393–408.

11. SALT CREEK

Knott, J. R., M. N. Machette, R. E. Klinger, and others. 2008. Reconstructing late Pliocene to middle Pleistocene Death Valley lakes and river systems as a test of pupfish (Cyprinodontidae) dispersal hypotheses. In *Late Cenozoic Drainage History of the Southwestern Great Basin and Lower Colorado River Region: Geologic and Biotic Perspectives*, eds. M. C. Reheis, R. Hershler, and D. M. Miller, Geological Society of America Special Papers 439, p. 1–26.

Wright, L. A., and B. W. Troxel. 1993. *Geologic Map of the Central and Northern Funeral Mountains and Adjacent Areas, Death Valley Region, Southern California.* USGS Miscellaneous Investigations Series Map I-2305.

12. TITUS CANYON

Miller, M., and L. A. Wright. 2015. *Geology of Death Valley National Park*, 3rd Edition. Dubuque, IA: Kendall/Hunt Publishing Company.

Niemi, N. A. 2012. *Geologic Map of the Central Grapevine Mountains, Inyo County, California, and Esmeralda and Nye Counties, Nevada.* Geological Society of America Digital Map and Chart Series 12.

Stock, C. 1936. Titanotheres from the Titus Canyon Formation, California. *Proceedings of the National Academy of Sciences* 22: 656–61.

13. UBEHEBE CRATER

Fierstein, J., and W. Hildreth. 2017. Eruptive history of the Ubehebe Crater cluster, Death Valley, California. *Journal of Volcanology and Geothermal Research* 335: 128–46.

14. RACETRACK PLAYA

McAllister, J. F., and A. F. Agnew. 1948. Playa scrapers and furrows on the Racetrack playa. Inyo County. California. *Geological Society of America Abstracts with Programs*: 1377.

Norris, R. D., J. M. Norris, R. D. Lorenz, J. Ray, and B. Jackson. 2014. Sliding rocks on Racetrack Playa, Death Valley National Park: first observation of rocks in motion. *PLOS ONE*, Public Library of Science 9 (8): e105948. Available online.

Reid Jr., J. B., E. P. Bucklin, L. Copenagle, and others. 1995. Sliding rocks at the Racetrack, Death Valley: What makes them move? *Geology* 23 (9): 819–22.

Sharp, R. P., and D. L. Carey. 1976. Sliding stones, Racetrack playa, California. *Geological Society of America Bulletin* 87: 1704–17.

Shelton, J. S. 1953. Can wind move rocks on Racetrack Playa? *Science* 117: 438–39.

Stanley, G. M. 1955. Origin of playa stone tracks, Racetrack Playa, Inyo County, California. *Geological Society of America Bulletin* 66: 1329–60.

15. RAINBOW CANYON AND FATHER CROWLEY VISTA POINT

Hall, W. E. and E. M. MacKevett. 1962. *Geology and Ore Deposits of the Darwin Quadrangle, Inyo County, California*. USGS Professional Paper 368.

Oswald, J. A., and S. G. Wesnousky. 2002. Neotectonics and Quaternary geology of the Hunter Mountain fault zone and Saline Valley region, southeastern California. *Geomorphology* 42 (3): 255–78.

16. THE MOJAVE RIVER

Cox, B. F., and J. W. Hillhouse. 2000. *Pliocene and Pleistocene Evolution of the Mojave River, and Associated Tectonic Development of the Transverse Ranges and Mojave Desert*. USGS Open-File Report 2000-147.

Enzel, Y., S. G. Wells, and N. Lancaster. 2003. Late Pleistocene lakes along the Mojave River, southeast California. In *Paleoenvironments and Paleohydrology of the Mojave and Southern Great Basin Deserts*, eds. Y. Enzel, S. G. Wells, and N. Lancaster, Geological Society of America Special Papers 368, p. 61–78.

Meek, N. 1989. Geomorphic and hydrologic implications of the rapid incision of Afton Canyon, Mojave Desert, California. *Geology* 17 (1): 7–10.

Reheis, M. C., and J. L. Redwine. 2008. Lake Manix shorelines and Afton Canyon terraces: Implications for incision of Afton Canyon. In *Late Cenozoic Drainage History of the Southwestern Great Basin and Lower Colorado River Region: Geologic and Biotic Perspectives*, eds. M. C. Reheis, R. Hershler, and D. M. Miller, Geological Society of America Special Papers 439, p. 227–59.

17. TRONA PINNACLES OF SEARLES LAKE

Phillips, F. M. 2008. Geological and hydrological history of the paleo–Owens River drainage since the late Miocene. In *Late Cenozoic Drainage History of the Southwestern Great Basin and Lower Colorado River Region: Geologic and Biotic Perspectives*, eds. M. C. Reheis, R. Hershler, and D. M. Miller, Geological Society of America Special Papers 439, p. 115–50.

Rieger, T. 1992. Calcareous tufa formations, Searles Lake and Mono Lake. *California Geology* 45 (4): 97–132.

Scholl, D. W., and W. H. Taft. 1964. Algae contributors to the formation of calcareous tufa, Mono Lake, California. *Journal of Sedimentary Petrology* 34 (2): 309–19.

18. FOSSIL FALLS

Blackwelder, E. 1954. Pleistocene lakes and drainage in the Mojave region, southern California. *California Division of Mines Bulletin* 170: 35–40.

Duffield, W. A., and C. R. Bacon. 1981. *Geologic Map of the Coso Volcanic Field and Adjacent Area, Inyo County, California.* US Geological Survey Miscellaneous Investigations Map 1200.

19. ANCIENT LAKE TECOPA

Hillhouse, J. W. 1993. *Late Tertiary and Quaternary Geology of the Tecopa Basin, Southeastern California.* US Geological Survey Map I-1728.

Reheis, M. C., J. Caskey, J. Bright, and others. 2020. Pleistocene lakes and paleohydrologic environments of the Tecopa basin, California: Constraints on the drainage integration of the Amargosa River. *Geological Society of America Bulletin* 132 (7/8): 1537–65.

20. THE RESTING SPRINGS PASS TUFF

Troxel, B. W., and E. Heydari. 1982. Basin and Range geology in a roadcut. In *Geology of Selected Areas in the San Bernardino Mountains, Western Mojave Desert, and Southern Great Basin, California*, compiler J. D. Cooper, Geological Society of America Cordilleran Section field trip guidebook, p. 91–96.

21. ALABAMA HILLS

Blackwelder, E. 1925. Exfoliation as a phase of rock weathering. *The Journal of Geology* 33: 793–806.

Stone, P., G. C. Dunne, J. G. Moore, and G. I. Smith. 2001. *Geologic Map of the Lone Pine 15' Quadrangle, Inyo County, California.* US Geological Survey Geologic Investigation Series Map I-2617.

22. LONE PINE FAULT AND THE 1872 EARTHQUAKE

Beanland, S., and M. M. Clark. 1994. *The Owens Valley Fault Zone, Eastern California, and Surface Faulting Associated with the 1872 Earthquake.* US Geological Survey Bulletin 1982.

Hauksson, E., B. Olson, A. Grant, and others. 2020. The normal faulting 2020 Mw 5.8 Lone Pine, eastern California, earthquake sequence. *Seismological Research Letters* 92 (2A): 679–98.

Hough, S. E., M. Page, L. Salditch, and others. 2021. Revisiting California's past great earthquakes and long term earthquake rate. *Bulletin of the Seismological Society of America* 111 (1): 356–70.

23. BIG PINE VOLCANIC FIELD

Vazquez, J. A., and J. M. Woolford. 2015. Late Pleistocene ages for the most recent volcanism and glacial-pluvial deposits at Big Pine volcanic field, California, USA, from cosmogenic ^{36}Cl dating. *Geochemistry, Geophysics, Geosystems* 16 (9): 2812–28.

24. POLETA FOLDS

Nelson, C. A. 1962. Lower Cambrian-Precambrian succession, White-Inyo Mountains, California. *Geological Society of America Bulletin* 73 (1): 139–144.

Nelson, C. A. 1966. *Geologic Map of the Blanco Mountain Quadrangle*. US Geological Survey map GQ-529.

25. ANCIENT BRISTLECONE PINE FOREST

Fenton, C. L. and M. A. Fenton. 1958. *The Fossil Book*. Garden City, NY: Doublday.

Nelson, C. A. 1966. *Geologic Map of the Blanco Mountain Quadrangle*. US Geological Survey map GQ-529.

Wright, R. D., and H. A. Mooney. 1965. Substrate-oriented distribution of bristlecone pine in the White Mountains of California. *The American Midland Naturalist* 73: 257–84.

26. BUTTERMILK BOULDERS

Bateman, P. C. 1965. *Geology and Tungsten Mineralization of the Bishop District, California*. US Geological Survey Professional Paper 470.

Furuki, H., and M. Chigira. 2019. Structural features and the evolutional mechanism of the basal shear zone of a rockslide. *Engineering Geology* 260: 105214.

27. OWENS RIVER GORGE

Hildreth, W., and C. J. N. Wilson. 2007. Compositional zoning of the Bishop Tuff. *The Journal of Petrology* 48 (5): 951–99.

Hildreth, W., and J. Fierstein. 2016. *Long Valley Caldera Lake and Reincision of Owens River Gorge*. US Geological Survey Scientific Investigations Report 2016-5120.

Wilson, C. J. N., and W. Hildreth. 2003. Assembling an Ignimbrite: Mechanical and Thermal Building Blocks in the Bishop Tuff, California. *The Journal of Geology* 111 (6): 653–70.

28. HILTON CREEK FAULT AT MCGEE CANYON

Berry, M. E. 1997. Geomorphic analysis of late Quaternary faulting on Hilton Creek, Round Valley and Coyote Warp faults, east-central Sierra Nevada, California, USA. *Geomorphology* 20: 177–95.

Sherburne, R. W., editor. 1980. *Mammoth Lakes, California, Earthquakes of May 1980*. California Division of Mines and Geology Special Report 150. See pages 49–73 on surface rupture and rockfalls.

29. HOT CREEK GEOLOGICAL SITE

Farrar, C. D., W. C. Evans, D. Y. Venezky, S. Hurwitz, and L. K. Oliver. 2007. *Boiling Water at Hot Creek—The Dangerous and Dynamic Thermal Springs in California's Long Valley Caldera*. US Geological Survey Fact Sheet 2007-3045.

30. MAMMOTH MOUNTAIN AND LONG VALLEY CALDERA

Bailey, R. A., G. B. Dalrymple, and M. A. Lanphere. 1976. Volcanism, structure, and geochronology of Long Valley caldera, Mono County, California. *Journal of Geophysical Research* 81: 725–44.

Hildreth, W. 2004. Volcanological perspectives on Long Valley, Mammoth Mountain, and Mono Craters: several contiguous but discrete systems. *Journal of Volcanology and Geothermal Research* 136: 169–98.

Hildreth, W., and J. Fierstein. 2017. *Geologic Field-Trip Guide to Long Valley Caldera, California*. US Geological Survey Scientific Investigations Report 2017-5022-L.

31. DEVILS POSTPILE

Beard, C. N. 1959. Quantitative study of columnar jointing. *Geological Society of America* 70, (3): 379-381.

Huber, N. K., and C. D. Rinehart. 1965. *Geologic Map of the Devils Postpile Quadrangle, Sierra Nevada, California*. US Geological Survey Geological Quadrangle Map GQ 437.

32. OBSIDIAN DOME, INYO CRATERS, AND THE EARTHQUAKE FAULT

Benioff, H., and B. Gutenberg. 1939. The Mammoth "Earthquake Fault" and related features in Mono County, California. *Bulletin of the Seismological Society of America* 29: 333–40.

Millar, C. I., J. King, R. Westfall, H. Alden, and D. Delany. 2006. Late Holocene forest dynamics, volcanism, and climate change at Whitewing Mountain and San Joaquin Ridge, Mono County, Sierra Nevada, CA, USA. *Quaternary Research* 66 (2): 273–87.

Reches, Z., and J. Fink. 1988. The mechanism of intrusion of the Inyo dike, Long Valley caldera, California. *Journal of Geophysical Research* 93: 4321–34.

Sieh, K., and M. Bursik. 1986. Most recent eruption of the Mono Craters, eastern central California. *Journal of Geophysical Research* 91 (B12): 12,539–571.

33. MONO CRATERS FROM PUNCH BOWL TO PANUM

Bevilacqua, A., M. Bursik, A. Patra, and others. 2018. Late Quaternary eruption record and probability of future volcanic eruptions in the Long Valley volcanic region (CA, USA). *Journal of Geophysical Research: Solid Earth* 123: 5466–94.

Hickman, K. T., and C. I. Millar. 2020. *Camera Trap Photographs from American Pika Haypiles in California*. US Forest Service Research Data Archive.

Putnam, W. C. 1938. The Mono Craters, California: *Geographical Review* 28 (1): 68–82.

Russell, I. C. 1889. Quaternary history of Mono Valley, California. *US Geological Survey 8th Annual Report*, p. 261–394.

Sieh, K., and M. Bursik. 1986. Most recent eruption of the Mono Craters, eastern central California. *Journal of Geophysical Research* 91 (B12): 12,539–571.

INDEX

Page numbers in bold include photographs.

aa, 88, 190, 191, **192**
Aberdeen, 186, 188
Aberdeen flows, 186, 188
Afton Canyon, 130, **134**, 135, 136
Aguereberry Point, 50
air-fall tuff, 165, 272
Alabama Hills, x, 169–76, **169**, **172**, 178, 180, 181, 185, 219, 225, 227
alluvial aprons, 13, 188, 284
alluvial fans, 10, **11**, 12–13, **14**, 31, 38, 39, 50, 73; groundwater in, 16; incised, **43**, 44; scarps cutting, **15**, 53, 179, 243. *See also* conglomerate; fanglomerate
alluvium, 22, 33, 44, 54, 56, 137, 146, 189, 190, 214, 242, 243
alteration, 36, 48, 173, **249**, 250, 259
Amargosa Canyon, 155
Amargosa River, 62, 63, 93, 130, 135, 137, 154, 155, 157, 161, 162
Ancient Bristlecone Pine Forest, 204–16
andesite, 32
anticlines, 5, 10, 83, 86, **97**
Apple Valley, 130
archaeocyathids, 207, **209**, 211
arches, 23, 176, 225
Argus Range, 121, 129
armoring, 225, **226**
Artist Drive Formation, 31, 32, **33**, **34**, 37, **40**, **48**
Artists Drive, 16, 19, 20, 21, 26, 29, 32–36, 52
Artists Palette, 35, **36**
ash, 7, 21, 155, 157–61, 165, 168, 230, 231, 254, 255, 271, 287
ash flows, 163, 164, 165
ash-flow tuff, **89**, 97, 165. *See also* tuff
Ashford Mill, 11, 58
Ash Meadows, 85, 155
Avawatz Mountains, 61

badlands, 16, 37, 154
Badwater, 9–15, 16, 39, 45, 46, 49, 59
Badwater Basin, 19, 25, **46**, **49**, 63, 86, 93
Badwater fan, **12**, 50
Badwater salt flat, 11, **14**, 16–19
Badwater turtleback, 10, 15, **52**, 57
bajadas, 13, **17**, 34
Baker, 63, 130, 131, 134, 135, 136, 154
Banner Peak, 263, 264
Barstow, 128, 130, 131, 133, 134, 135, 164
basalt, 3; cinders of, 103; cobbles of, 34; columnar joints in, 263, 265–69; dikes of, 32, 121, **122**, **126**; flows of, 32, 37, 121, 123, 124, **125**, **126**, 127, **128**, 129, **149**, **150**, 187–93, **192**, **193**, 237; potholes in, **151**, **152**; strandlines in, 59; tephra, **105**, 114, **123**; ventifacts of, **28**, **29**. *See also* aa; cinder cones; pahoehoe
basalt east of Little Lake, 146, 149, **150**, 151
basalt of Red Hill, 146, **149**, **150**
base surge, 107
Basin and Range, x, 1, 2, 4, 5, 7, 14, 15, 121, 177, 240, 278
batholiths, x, 7, 124, 129, 258
beach ridges, 58, **61**, 88, 135, 139
Beatty, 88, 93
Beatty Cutoff Road, 58, **61**, 84, 88
Beatty Junction, 84, 86, 88
bedding, 79, 80, 81, 86, 94, 100, 166, 199, 200, **202**, 209
beryllium, 141
Big Pine, 7, 176, 182, 184, 186, 187, 188, 195, 255
Big Pine volcanic field, 4, 182, 186–93
Billie Mine, **48**
biotite, 219, 234, 259

306

INDEX   307

Bishop, 158, 183, 194, 211, 217, 219, 229, 232, 257
Bishop ash, 158, 160, **161**
Bishop Creek, 8, 211, 218, 263
Bishop Tuff, x, 160, 164, 166, 187, 229–38, **234**, **235**, **236**, **238**, 255, 260
Black Mountains, x, 15, **20**, **33**, **40**, 49–53, 56; alluvial fans of, **12**, 13, **35**, 44; rocks of, 32, 33; scarps along, 21, **35**, 54, 55, 60, 83; strandlines along, **11**, 58, 60; turtlebacks of, 52–56
Black Mountains volcanic field, 21
Blanco Mountain, 212
Bonanza King Formation, 48, **50**, 90, 93, 94, 96, 98, **99**, 100, **101**, **128**, **163**
Borah Peak, 242
boron, 140, 141
breccia, **76**, 77, **79**, 81, 82, 88, 100, 101, **102**, 114, 163. *See also* megabreccia
bristlecone pines, 205, **206**, 211–16, **213**, **215**
Bristol Lake playa, 134
Bristol Mountains, **134**
Bullfrog Mine, 91
Buttermilk Boulders, 217–27

calcite, 73, 93, 101, 121, 213, 249, 250, 259
calcium carbonate, **142**, 143
calderas, 7, 92, 97, 158, 255, 263. *See also* Long Valley caldera
caliche, 121
California Volcano Observatory, 243, 271
Cambrian, x, 6, 47, 48, 89, 90, 96, 99, 163, 197, 207, 210
Camp Independence, 183
Campito Formation, 207, 214
carbonate rock, 44, 73, 216. *See also* dolostone; limestone
carbon dioxide, 7, 143, 261
Carrara Formation, **50**, 90, 99, **100**
Casa Diablo Hot Springs, 247, 250, **251**, 252, 262

Castle in Clay, 154, 161
Catcher's Mitt, 222, 223, **224**, 225, 226, 227
Cedar Flat, 204, 207, 208, 209
Charleston Peak, 63
chert, 97, 98
China, Lake, 62, 141, 147
cinder cones, 103, 122, 123, 124, **127**, 147, 148, **149**, 186, 187, 188, **189**, 190, **191**
clay, 16, 36, 83, 84, **107**, 155, 174, 175, **249**, 250, 258, 259
claystone, 86
Clem Nelson Peak, 194, 208, **212**
Colorado Plateau, 1, 2
Colorado River, 23, 42, 84, 134
columnar jointing, 149, 167, 191, **235**, **236**, 263, **264**, **266**, 268, 269
conglomerate, 86, 103, 106, 155, **166**; of alluvial fan, 21, 37, 39, 40; of Mormon Point Formation, 21, 24, 25, 33; of Titus Canyon Formation, 89, 90, 92, 93, **94**, 95, 96, 97, **98**
Convict Creek Canyon, 244
Convict Lake, 241, 249
Copper Canyon fan, 10, 18
Copper Canyon turtleback, 10, 54, **55**, 57
core stones, 219, 227
Corkscrew Peak, 61
Coso Hot Springs, 147
Coso Junction, 146, 147
Coso Range, 7, 147, 148, 153
Coso volcanic field, 4, 147, 148, 149, 187
Cottonball Basin, 83
Cottonball Marsh, 85, 86
Cottonwood Mountains, 65, 103, 114, 129
coulees, 272, 281, 282, 283, 284, 287, **288**, 289
craters. *See* Crater Mountain; Inyo Craters; Little Hebe Crater; Mono Craters; Panum Crater; Pisgah Crater; Ubehebe Crater
Crater Mountain, 184, 186, 187, **188**, 190, 282, 284
creosote bush, **67**

cross bedding, 61, 100
Crowley, Father, 128
Crowley, Lake, 128, 232, 239, 240, 254, 255

dacite, 254, 259
Dantes View, 10, 13, 15, 19, 45–56, 58, 59
Darwin Plateau, 122, 126, 129
Darwin Wash, 121
dating, isotopic, 206, 268, 287
Daylight Pass, 58, 61, 88
Deadman Dome, 270, 278, 281
Deadman Summit, 271
Death Valley, x, 1, 6, 13, **49**, 51
Death Valley fault zone, 51, 83, 84, 102
Death Valley National Park, 10, 24, 119, 121
debris flows, 8, 9, 15, 18, 22, 34, 55, 73–80, 102, 173, 175, 180, **181**, 183
Deep Spring Formation, 99, 212, 214, 216
Deep Springs Valley, 194, 195, 196, 208
Deer Mountain, 270, 275, 276, **277**
deformation, 5, 25, 100, 106
desert varnish, **153**, **156**, 172, **175**, **202**, 207
detachment faults, 3, **24**, 25, 53, 55, 56
Devils Golf Course, 9, 10, 16–20, **17**
Devils Hole, 85
Devils Playground, 134, 135
Devils Postpile, x, 258, 259, 262, 263–69, **264**, **266**
Devonian, 47
Diaz Lake, 182, 184
dikes: basalt, 32, 121, **122**; feeder, 270, 274, 275, 276, 277, 278, 280, 284, 287; granitic, 259; Independence, 123, **124**, **126**; Jurassic, 123, **124**, **126**
dip-slip faults, 5
Discovery Trail, 206, 212, 213, 216
Division Creek Rest Area, 186
dolomite, 73, 213

dolostone, 73, 90, **101**, 111, **112**, 114, 115, **116**. *See also* Bonanza King Formation; Reed Formation
Dry Creek Dome, 276
Dublin Gulch, 154, 158
dunes, 64–71, 125, 129

Eagle Mountain, **156**
earthquakes, 144, 145, 171, 177–85, 219, 227, 240–45, 254, 261; boulders dislodged by, 227, **244**, **245**; sediment deformed by, **160**, 161
Earthquake Dome, 270, 276
Earthquake Fault, 271–80, **279**, 284, 287
Eocene, 89, 90, 93
epicenters, 240, 241, 254
erosion: by diversions, **41**, **42**, **43**; by glacial floods, 148, **151**, **152**; headward, 41; resistance to, 232; scour and fill, 73, 75, 76, 79; old surface of, 126, 258; of surficial deposits, 103, **108**; by wind, 26–31, 231. *See also* piping; sandblasting
erratics, 219, 222, 250
Eureka Dunes, 65

Fall Canyon, 102
fanglomerate, 21, 22, 24, 33, 35, **43**, 55, **56**, 121, **122**, 123, 206
fans. *See* alluvial fans
Father Crowley Vista Point, 120–29
faults, 4–5, **24**, **55**, **56**, **97**, 105, 126, **128**, **166**, **198**. *See also* detachment faults; normal faults; scarps; strike-slip faults; thrust fault; *specific fault names*
feldspar, 36, 111, 166, 174, 175, 178, 234, 249, 259, 273
Fish Lake Valley, 102
Fish Springs Hill cone, 186, 187, **189**, 190
Fit Homeless, 220, 226, 227
flash floods, 8, 9, 21, 72, 75, 78, 79, 88, 101, 102
flow banding, 249, 250
folds, 4–5, 80–81, **80**, **81**, **97**, **128**, **198**, **200**. *See also* Poleta folds

fossils, 47, 93, 95, 99, 157, **203**, 204, 207, 209, 210
Fossil Falls, 146–53, **148**, **150**, 229, 257
fumaroles, 231, **232**, 236, 249, 250, **251**, 252, 258, 259, 261
Funeral Formation, 33, 86
Funeral Mountains, 44, **48**, 50, **52**, 93
Funeral Peak, 49
Furnace Creek (creek), 37, 38, 40, 41, 63
Furnace Creek (town), 10, 14, 27, 39, 52, 59
Furnace Creek fan, 10, 38, 44
Furnace Creek Formation, 32, **33**, 37, 40, 42, 48, 84, 86
Furnace Creek Wash, **3**, 37, 38, 39, 40, 41, 42, 44, 57

geologic time, x, 5–6
geothermal, 147, 247, 248, 249, 250, 251, 258, 274
Glacial Owens River, 146, 147–49, 153, 230
glaciation, x, 1, 8, 59, 139, 181, 267, 268, 269
glaciers, 8, 57, 62, 147, 181, 205, 222, 233, 237, 239, 241, 258, 269; polish by, **267**
glass, **164**, **167**, 168
Glass Creek, 272, 274
Glass Creek Flow, 259, 270, 272, 274, 281
Glass Mountain, x, 230, 246, 254, **255**, 258, **259**, 260
gneiss, 24, 25, 33, 49, 52, 55, **56**
gold, 38, 91, 222, 258
Golden Canyon, 10, 33, 37–44, **40**, 53
Golden Canyon fan, 38
Gold Mountain, 258
Goodale Campground, 190
Goodale Canyon, 188
Goodale Creek, 186, 190
Gower Gulch, 10, 33, 37–44, **40**
Gower Gulch fan, 33, 38, 39, **40**, **43**
Grand Canyon, 42, 47, 48, 52, 164
Grandma Peabody, 217, 220
Grandpa Peabody, 217, 218, 219, **220**, **221**, 222, **223**, 227

Grandstand, 110, 111, 114
Grandview Campground, 194, 204, 209
granite, 124, 131, **132**, 133, **221**, 222, **264**; armoring of, **226**; boulders of, 218–27; cobbles of, 177, 178, **180**, 284, 286; crystals in, 111, 178; dikes of, 259; joints in, 173, **174**; steptoes of, 187, **188**; Triassic, 237, **238**; weathered, 65, 169, **172**, 174, **175**; xenoliths of, 189. *See also* Alabama Hills; Buttermilk Boulders; Hunter Mountain batholith; Sierra Nevada batholith; Spangler Hills
Granite Mountains, 173
granodiorite, 121, 122, 124, **125**, 126, 127, 129, 187, 188
Grapevine Mountains, 52, 83, 89, **90**, 92, 93, 101, 105
gravel; cemented, **76**, **79**, **82**; of delta, 286; in fan deposits, 12, 74; in shoreline deposits, **11**, 14, 35, 57, 58, 60, 139; surface, **156**; in wash, 41, 44, 73, 75, 77, **78**. *See also* conglomerate
Green Wall, **217**, 220, 225, 226
Grotto Canyon, 67, 68
Grotto Canyon fan, 64, 66
groundwater, 16, 24, 39, 63, 83, 123, 132, 133, 137, 143, 184, 195; and magma, 103, 275, 276

Haiwee Reservoir, 147
Hanaupah fan, 10, **12**, 50
hanging chutes, **22**
Happy Boulders, 217
Harkless Formation, 197, 198, 202
Hartley Springs Campground, 270, 271
Hartley Springs fault, 240, 243, 254
Hill 1660, 190, **191**, 192
Hilton Creek fault, 239–45
Horseshoe Lake, 253, 256, 260, 261, 276
Horseshoe Meadows, 171, 172
Hot Creek, 141, 243, 245, 246, **248**, **255**, 258
Hot Creek Geological Site, 246–51, **247**, **248**

Hot Creek Trout Hatchery, 245, 246, 247
hot springs, 141, 147, 154, 155, 156, 157, 246, 247, 250, 251, 252, 261, 262, 284
Huckleberry Ridge ash, 158
Hunter Mountain, 110, 127
Hunter Mountain batholith, 124, 129
Hunter Mountain fault, 127, 129
Huntley Clay Mine, 258
hydrothermal alteration, 36, 48

Ibex Hills, **156**
Ibex Pass, 154, 157
ice age, 57, 147. *See also* glaciation; Pleistocene
igneous, 6, 45
Independence, 182, 183, 184
Independence dike swarm, 123, **124**, **126**
Indian Wells Valley, 147
Inyo County, 49, 128
Inyo Craters, 103, 254, 255, 259, 270–80, **277**, 284, 287
Inyo dike, 270, 277, 278
Inyo Domes, 254
Inyo Mountains, 124, 129, 169, 171, 176, 180, 183, 195, 206
Inyo volcanic chain, 271, **273**, 276, 280
iron, **24**, 25, 36, 122, 222, 227
isoclinal folds, 5, 81
isotopic dating, 206, 268, 287

Johnnie Formation, 81, 90
joints, 23, 173, 174, 191, 236, 263, 269
June Lake Junction, 270, 271, 283, 287, 288
Jurassic, 121, 123, 124, 126

Kaibab Limestone, 47
kaolinite, 250, 258, 259
Kelso Dunes, 65
Kern River, 263
Keyhole unit, **74**, 75, 76
Klare Spring, 99, 100

Lahontan, Lake, 84
lakebeds, 39, 130, **141**, 154, 155, **156**, 157, 158, 162, 163, 230, 254
Last Glacial Maximum, 8
Lava Creek ash, 158, **159**, **160**
lava flows, 7, 32, 37, **89**, 104, **125**, **149**, **150**, **188**, 187–93; columnar joints in, 265–66; 268–69; scarps in, 185; water blocked by, 134. *See also* aa; basalt flows; pahoehoe; rhyolite flows
lava tubes, **193**
Leadfield, 88, 93, 95, 96, 97, 98, 99, 100
Lee Vining Canyon, **288**
left-slip faults, 4, 5, 198. *See also* strike-slip faults
limestone: cobbles of, 44, 106; Paleozoic, 47, 90, 123, 124, **126**, **163**, 214, 216, 259; of Poleta Formation, 197, **198**, **201**, **202**, 203, 207, **209**; Proterozoic, 46, 73, 90. *See also* marble
lithium, 141, 195
Little Hebe Crater, 104, **107**, **108**, 109
Little Lake, 146, 147, 148, 149, 150, 151
Little Poleta, 196, 197, 198, 203
Lone Pine, 147, 170, 171, 176, 177, 178, 179, 183, 185
Lone Pine earthquake, 185
Lone Pine fault, 171, 177–85
Long Valley caldera, x, 160, 229–32, 237, 253–61, **255**, **259**; faults in, 239, 240, 241, 243
Long Valley Dam, 128
Long Valley Lake, 62
Los Angeles Aqueduct, 170, 171, 178, 179, 229, 232

magma: baking by, 121; and calderas, 230, 231, 243, 253; groundwater and, 103, 284; heat from, 246; rhyolitic, 271, 272, 273, 274, 276, 281, 287; subduction and, 7, 124, 141; xenoliths and, 189, 234
Mammoth Creek, 246
Mammoth Crest, 258

Mammoth Knolls Dome, 270, 276
Mammoth Lakes (town), 242, 246, 252, 253, 255, **256**, 258, **259**, 262, 271, 277, 280
Mammoth Lakes Airport, 245
Mammoth Lakes Basin, 246, **256**, 258
Mammoth Mountain, 187, 230, 246, 252, 253–61, 262, 271, **273**, 278
Mammoth Pass, 259
Mammoth Rock, 252, 253, **259**, **260**
Mammoth School, 245
Mammoth Yosemite Airport, 247
Manix, Lake, 62, 130, 133, 134, 135, 139, 158
Manly, Lake, 57–63, 84–85, 130, 137, 147, 155, 158; beach ridges of, **61**, 88; deposits of, 33; shorelines of, 11, 14, 35, 57, **60**
Manly Beacon, 37
mantle, 65, 67, 106, 107, 142, 143, 187, 189, 286
marble, 54, 55, 73, 76, **77**, 78, **79**, **80**, 81, 82, 131, 259, 260
Mars, 26, **28**
Mars Hill, 26
Mary, Lake, 246, 252, 253, 262
McCoy Station, 252, 256
McGee Canyon, 239–45, **241**, **243**, **255**
McGee Creek Campground, 239, **244**
McGee Mountain, 239, 242, **243**, 244
McLean Spring, 86
megabreccia, **92**, 93, **94**, **96**, 114
Mesa Falls ash, 160
mesquite, 39, 65, **67**
Mesquite Basin, 83
Mesquite Dunes, 64, 65–71, **66**, **70**, **71**, 83, 86
Mesquite Flat, 63, 65, 68
metamorphic rock, 49, 53, 54, 222, 243, 259, 260, 284, 286
Methuselah Trail, 212, 214
mica, 234, 259
Middle Fork of the San Joaquin, 258, 262, 263
Mill Creek, 228
Minaret Summit, 259, 262
Minaret Vista, 262, 263, 264, **265**
Mine Road Dome, 282

Miocene, 32, 33, 37, 48, 49, 55, 86, 89, 90, 92, 93, 95, 106, 125, 163
Mojave, Lake, 62, 130, 136, 137
Mojave Desert, 36, 47, 65, 85
Mojave Narrows, **132**, 133, 135
Mojave River, 62, 63, 130–37, **132**, **133**, **134**
Mono Basin, 281, 284
Mono Craters, 4, 187, 254, 256, 281–89, **283**
Mono Lake, 143, 230, 254, 255, 259, 271, 281, 284, 286, 287
Montgomery Mine, 91
Montgomery Peak, 238
moraines, 8, 218, 219, 233, 239, 240, **241**, **243**, 269
Mormon Point, 10, 18, 31, 50, 55, 56, 60
Mormon Point Formation, 21, 24, 25, 33
Mormon Point turtleback, 10, 55, **56**
Mosaic Canyon, 72–82, **77**, **78**, **79**, **82**
Mosaic Canyon fan, 72
Mosaic Canyon fault zone, 72, 81
movies, 139, 170, 172
mud cracks, 38, **68**, 114, **115**
mudstone, 32, 33, 89, 90, 93, 106, 155, **156**, 158
Mushroom Rock, 32, 38

Native Americans, 131, 137
Natural Bridge, 10, 21, **23**, 24
Natural Bridge Canyon, 4, 15, 21–25, 54, 55, 56
Naval Air Weapons Station China Lake, 147
Nelson, Clem, 197, 206, 207–8, **207**, 212, 244
Nevares Springs, 63
Noonday Formation, 73, 75, **76**, 77, **79**, **80**, 81, 82, 90
Nopah Range, 50, **163**
normal faults, 3, 4, 5, 7, 15, 51, 126, 166, 176, 177, 233, 239, 240, 254, 277
North American Plate, 7
North Coulee, 282, 287, 288
North Palisade, 49

North Pinnacles, 142
Northwest Coulee, 282, 287, 288

obsidian, 168, 258, 271, 272, 273–74, **274**, **275**, 284, 286
Obsidian Dome, 187, 249, 255, 270–80, **273**, **274**, 281, 284
Olancha, 182
Oligocene, 89, 90, 93, 95
olivine, 36, 127, 150, 189, 190, 265
Ordovician, 47
oreodonts, 95
Oro Grande, 130, 131, 133
Owens Lake, 62, 63, 147, 183
Owens River, 62, 63, 84, 128, 139, 140, 229, 232, 233, 237, 238; Glacial, 146, 147–49, 153, 230
Owens River Gorge, 166, 173, 228–38, **232**, 244, 255
Owens Valley, 1, 4, 13, 128, 169, 176, 177, 185, 207; ancestral, 230, 231; glacial meltwater in, 147, 148
Owens Valley earthquake, 178, 182, 183
Owens Valley fault, 176, 178, 179, 182, 183, 184, 185, 186, 188, 189

Pacific Ocean, 46, 97, 131
Pacific Plate, 7
pahoehoe, 190, **192**
Pahrump, Lake, 62, 130
Paleozoic, 6, 45, 47, 48, 50, 51, 52, 90, 92, 95, 96, 124, 127, 128
Palmer, Mt., 88, **90**, **96**, 102
Panamint, Lake, 62, 121, 130, 139, 147
Panamint Butte, 120, 125, 127, 128, 129
Panamint Range, 32, 37, **43**, 45, **46**, 49, 51, 52, 53, 54, 56, 67, 127, 129, 177; alluvial fans of, **11**, 12, 13, **17**, **30**, 31, 50
Panamint Springs, 120, 121, 122, 124, 125
Panamint Valley, x, 1, 120, 121, 124, 125, 126, 127, 129, 138, 139
Panuga Formation, 90
Panum Crater, 249, 281, 282, **283**, **286**, 287, 288

Paoha Island, 254, 271, 284
Payson Canyon, 195
Peabody Boulders, 219, **220**, **221**, 222, 225, 227
Peach Springs Tuff, 164
Pearsonville, 146, 147
Pennsylvanian, 123
peridotite, 189
Permian, 47, 123
petroglyphs, 147, **153**
Pine Creek, 8, 170, 171, 178, 228, 233
pinnacles, 139–45, **140**, **144**, **145**
Pinus longaeva, 205
Pinyon Picnic Area, 209
piping, 24
Pisgah Crater, 134
Pisgah fault, 135
Pisgah lavas, 130, 134
plagioclase, 127, 150, 190, 219, 265
playas, 8, 62, 67, 111, 114–15, 125, **136**. See also Bristol Lake; Racetrack; Searles Lake; Silurian Lake; Silver Lake; Soda Lake
Pleistocene ice ages, x, 14, 25, 57, 62, 136, 138, 139, 147, 155, 186, 230, 289. See also Tahoe glaciation; Tioga glaciation
Pliocene, 32, 33, 86, 90, 122, 123, 126
pluvial lakes, 8, 11, 14, 61, 62, 63, 84, 85, 121, 139, 140, 147, 284
Poison Canyon, 141
Poleta folds, 195–203, **196**, 208
Poleta Formation, 197, 198, 202, 204, 207, 209, 213
potholes, 146, **151**, **152**, 153
Proterozoic, x, 6, **24**, 25, 45, 49, 52, 54, **56**, 73, 81, 89, 99, 213
Protitanops curryi, 95
pull-apart basins, 51, 52
pumice, 165, 166, **167**, 168, 230, 231, 233–35, 273, **275**, 284; insulating properties of, 289; mining of, 157; miring by, 272, 283; ring of, 284, **286**, 287
Pumice Mine, 282, 283
Pumice Pit Dome, 282
Pumice Valley, 284

INDEX   313

Punch Bowl, 281, 282, 283, 284, **285**, 286, 287, 288, 289
pupfish, **83**, 84, 85, 86, 87
pyrite, 222
pyroclastic, 168, 230, 231, 232

quartz, 65, 166, 173, 174, 200, 219, 220, 222, 233, 259, 273
quartzite, 47, 90, 97, 98, 99, 100, 106, 133, 189
Quartzite Mountain, 133

Racetrack, 111–19, **115**, **116**, **117**, **118**
Racetrack Valley, 111, 114
Rainbow Basin, 36
Rainbow Canyon, 120–29
Rainbow Falls, 262
Red Cathedral, **33**, 37
Red Cliffs, 43
Red Hill, 146, **149**, 150
Red Pass, 88, 93, 95, 96
Reds Meadow, 262
Red Wall Canyon, 102
Reed Flat, 211, 212
Reed Formation, 208, 212, 213, 214, 216
Resting Spring Pass, 157, 162, 164
Resting Spring Pass Tuff, 93, 162–68, **164**, **166**, **167**
Resting Spring Range, 50, 154, **156**, **163**
resurgent domes, 240, 254
reverse faults, 4, 5
rhyolite, 272–73, **289**; domes of, 147, 187, 230, 245, 246, 247, 249, 250, 258, 272, **283**, **285**, **286**; flow-banded, 249; flows of, 92, 186, 237, 247, 272, 277, 278, 287, 288; magma, 103, 164, 255, 273; ventifact of, **29**
Rhyolite (ghost town), 88, **91**, 93
Ridgecrest, 130, 139
Ridgecrest earthquake, 144, 145
right-slip faults, 4, 51, 52, 129, 181, 183. *See also* strike-slip faults
ripple marks, 38, 65, **70**, **71**, 100, **200**
Ritter Range, 256, 258, 263

rock art, 147, **153**
Rock Creek Gorge, 237
Rodinia, 6, 45, 46
Rose Valley, 147
Rush Creek, 286, 287
Russell, Lake, 62, 284

sag ponds, **184**
Saline Valley, 1, 62, 121, 129
salt, 9, 16–20, **17**, **20**, 86, 140; saucers of, 19–20; wedging by, 18
Salt Creek, 16, 19, 20, 63, 83–87, **85**
Salt Creek anticline, 83, 86
Salt Creek fault, 83, 86
Salt Creek Hills, 83, 86
salt pan, 9, **11**, 13, **14**, 15, 16, 17, 18, 19, 45, 50
Salt Spring Hills, 137
San Andreas fault, 7, 132, 177
San Bernardino Mountains, 49, 62, 63, 130, 131, 132, 135
sandblasting, 26, 28, 29, 30, 31
sand dunes, 64–71, 125, 129
sandstone, 46, 47, 93, **98**, 155, **166**, 197, **198**, **200**, **201**, **202**. *See also* Deep Spring Formation; Poleta Formation; Titus Canyon Formation
San Gabriel Mountains, 242
San Joaquin River, 252, 255, 258, 259, 262, 263
Saratoga Springs, 85
scarps, **15**, 21, **23**, **33**, **35**, 39, 53, **54**, 60, 102, **188**, **189**, 277; of Hilton Creek fault, 239–45, **241**, **242**; of Lone Pine fault, 170, 171, 177–85, **179**; Sierran, 176, 177, 209, 239
schist, 18
Schulman Grove, 204, 206, 208, 211, 212, 214
Searles, Lake (pluvial), 8, 62, 139, 140, 141, 142, 143, 144, 147
Searles Lake playa, 138–45
Searles Valley Minerals Inc, 140
sedimentary rock, 1–3, 7
sedimentation, x, 47, 124, 237
shale, 46, 47, 90, 159, 196, 197, 198, 202, 209
Sharma, **225**, 227

Sherwin Grade, 229
shorelines, **11**, 14, 35, 39, 47, 57, 58, 59, **60**, 135, **136**, 137, 138, 139, 142
Shoreline Butte, 10, 52, 55, 58, 59, **60**
Shoshone, 154, 155, 158, 159, 160, 161, 162
Shoshone Mine, 91
Shoshone Museum, 154, 161
Sierra Nevada, 1, 2, 7, 49, **169**, 188, 205, **210–11**, **233**; escarpment of, 176, 177, 209, 239; fault along base of, 171, 177, 182, 183, 239; glaciation in, 8, 59, 62, 63, 139, 147, 148, 181, 269
Sierra Nevada batholith, x, 7, 124, 258
Sierra View, 194, 204, 209, 211
Sierra wave, **257**
silica, 97, 247, 250
siliceous, 250
silt, 16, 24, 30, 31, 67, 73, 78, 136, 155, 198
siltstone, 32, 33, 37, 39, 40, 105, 123, **124**, 202, 207, 259
Silurian, 47, 90
Silurian Lake, 135, 137
Silver Lake, 63, 130, 131, 135, **136**, 137
slot canyons, 1, 42, 73, 77, 88, 102
Soda Lake, 130, 135, 136
soil, **125**, 212, 213, 214
South Coulee, 282, 287
Spangler Hills, 138, 142, 143
spatter, 109, 191
spheroidal weathering, 173
Split Mountain, **49**, **210**
Spring Mountains, 63, 155
star dunes, 64, 68
steptoes, 187, **188**
St. Helens, Mt., 160
Stirling Formation, 90
Stovepipe Wells, 64, 65, 72, 73, 88
strandlines, 57, 59, **60**, 136, 138, 142. *See also* shorelines
stratigraphy, 90, 197
strike-slip faults, 4, 51, 52, 53, 129, 176, 177, 181, 183, 198
synclines, 5, 98, 199, **200**, 201, 202

Taboose Creek, 190
Tahoe glaciation, 8, 59, 138, 139, 142, 143, 239, 269
talus, 237, 264, 265, 283, 288
Tapeats Sandstone, 47
Taphrhelminthopsis, **203**
Tea House, **3**
Teakettle Junction, 110, 114
Tecopa, Lake, x, 62, 130, 155–61, 162, 164, 230, 254
Tecopa Basin, 154, **156**
Tecopa Hot Springs, 154, 155, 156, 157
tectonics, 5, 7, 51, 56, 81, 83, 168, 177, 278
Telescope Peak, **30**, 31, 37, **43**, **46**, **49**, 58
tephra, 7, 103, **105**, **106**, **108**, 109, **123**, 187, 189, 229, 230, 231, 255, 272, 285
Texas Springs, 63
Thimble Peak, 88, **92**
thrust faults, 5, 7, 128, 196
Tinemaha Reservoir, 186
Tin Mountain, 110, 114
Tioga glaciation, 8, 59, 138, 139, 142, 143, 148, 181, 239, 267, 269
Tioga Pass, 211
titanotheres, 95
Titanothere Canyon, 88, 93, 94
Titus Canyon, 5, 88–102, **101**, **102**
Titus Canyon fault, 99, 100
Titus Canyon Formation, **89**, 90, 92, 93, **94**, 95, **96**, **97**, **98**
Toll House Spring, 207
Tom, Mt., **211**, 221, **222**, **233**
Toms Place, 240
trace fossils, **203**
Trail Canyon, 50
Trail Canyon fan, **46**, 50
transverse dunes, 64, 67, 69, 70, 71
travertine, 250
Travertine Springs, 38, 63
Triassic, 124, 237, 238
trilobites, 207, 210
Trona, 138, 139, 140
Trona Pinnacles, 139–45, **140**, **144**, **145**, 153

INDEX 315

Tucki Mountain, 37, 64, 68, 73, 81, 82
tufa, 1, 57, 121, 139, **142**, 143, 144, 155, 259
tuff, 32, **34**, **36**, **164**, 165, **167**, **234**; mudstones rich in, **156**. *See also* air-fall tuff; ash-flow tuff; Bishop Tuff; Resting Spring Pass Tuff
tumuli, **191**, 192
Tungsten Hills, 218, **233**
Tuolumne Meadows, 178
turtlebacks, 15, 45, 52, 53, 54, 55, 56, 57
Tuttle Creek, 170, 171, 173, 175
Twin Lakes, 246, 252

Ubehebe Crater, x, 4, 7, 103–9, **104**, **107**, 114
Ubehebe Peak, 110
Ubehebe volcanic field, 103, 106
unconformities, 33, 47, 90
Union Pacific Railroad, 135
Upper Gorge Power Plant, 228, 233

varnish, desert, **153**, **156**, 172, **175**, **202**, 207
veins, 190
vents, 103, 109, 149, 150, 165, 187, 268, 272, 281, 284, 286
ventifacts, 26, **28**, **29**, **30**, 31
Ventifact Ridge, 10, 26–31, **27**
vesicles, 26, 29, 165, 166, 168, **193**
Victorville, 130, 131, 132, 133, 135
volcanic ash, 1, 7, 33, 155, 157, 163, 164, 230, 254, 256
volcanic bomb, 7, **109**, 189, **190**
Volcanic Tableland, 229, 230, 231, **233**, 238

water table, 9, 278
weathering, 123, 124, 171; cavernous, 34; differential, 235; of granite, 65, 172, 173, 174–75, 219, 222–24, 226, 227; pits, **223**, 224, **225**, 227; salt, 10, 18; spheroidal, 173. *See also* erosion; hydrothermal alteration
welded tuff, 92, 161, 163, 164, 165, **167**, 231, **234**, 237
Western Mining Company, 96
Westgard Pass, 194, 195, 204, 206, 209
Wheeler Crest, 211, 233
White-Inyo Range, 47, 197, 207, 208
White Mountain Peak, 232, 246, 259
White Mountains, 99, 204, 205, 206, 208, 210, 213, 230, **238**, 246, 259
White Pass, 92, 93, 95
Whitney, Mt., 14, 49, 128, 169, 172, 176, 177, 178, 182, 184, **185**, 211, 229
Williamson, Mt., 49
Wilson Butte, 270
wind ripples, 70
Wood Canyon Formation, 90, 99, 100

xenolith, 189, 190

Yellowstone, 154, 158, 159, 160, 164, 231
Yosemite National Park, 173, 178, 263, 284, 288
Yosemite Valley, 183

Zabriskie Point, 10, 33, 37–44, 48
Zabriskie Quartzite, 90, 99, **100**

Art (left) and Allen (right) sitting on upside-down Zabriskie Quartzite in Titus Canyon.

Allen F. Glazner and **Arthur Gibbs Sylvester** were born, raised, and educated in southern California. They each earned BA degrees from Pomona College and PhD degrees in geology from UCLA. **Allen** was a sports reporter in high school, worked summer jobs with the USGS and Shell, and took a faculty position at the University of North Carolina at Chapel Hill immediately after finishing his PhD, pursuits that were all too fun to be considered real jobs. He has conducted geological research in the Sierra Nevada and eastern California since his undergraduate days, and retired from teaching in 2019. He is coauthor of *Geology Underfoot in Yosemite National Park* and of both editions of *Geology Underfoot in Southern California*, the latter with Art and Bob Sharp. **Art** joined a team of Shell Development Company research geologists to study the tectonic history of the Pacific margin of the United States after earning his PhD, but UC Santa Barbara lured him away from Shell to teach courses in structural geology, field geology, and petrology. His research includes fieldwork in Norway, Italy, Grand Teton National Park, the Tahoe-Sierra, and the Salton Trough. He has led more than 300 field trips in southern California for student, industrial, and professional geologists. After retiring from active teaching in 2003, Art and Libby O'Black Gans coauthored *Roadside Geology of Southern California*.

Robert P. Sharp (1911–2004) originated the *Geology Underfoot* series. In 1993 he took on Allen, 43 years his junior, to cowrite the first book in the series, *Geology Underfoot in Southern California*. Bob taught generations of geology students at the California Institute of Technology (Caltech), leading them on field trips across the West and Hawaii even into his 90s. A true polymath who could explain complex concepts to anyone, his specialty fields included sand dune physics, glaciology, and planetary sciences. He was among the first to interpret dune processes in orbital imagery from Mars and received the Penrose Medal from the Geological Society of America, its highest honor, in 1977. In further recognition of his contributions to diverse fields, a mountain and a glacier in Antarctica, a Mars-crossing asteroid, and a mountain on Mars have been named in his honor.

GEOLOGICAL CHECKLIST FOR EASTERN CALIFORNIA

- [ ] aa lava
- [ ] air-fall tuff
- [ ] alluvial fan
- [ ] anticline
- [ ] ash-flow tuff
- [ ] bajada
- [ ] basalt
- [ ] beach ridge
- [ ] Bishop Tuff
- [ ] Bonanza King Formation
- [ ] breccia (sedimentary)
- [ ] breccia (tectonic)
- [ ] bristlecone pine
- [ ] caldera
- [ ] cinder
- [ ] cinder cone
- [ ] columnar joints
- [ ] conglomerate
- [ ] coulee
- [ ] dacite
- [ ] debris flow
- [ ] detachment fault
- [ ] dike
- [ ] dip-slip fault
- [ ] dolostone
- [ ] fanglomerate
- [ ] fault scarp
- [ ] fossil organism
- [ ] fumarole
- [ ] gneiss
- [ ] granite/granodiorite
- [ ] halite
- [ ] hot spring
- [ ] isoclinal fold
- [ ] joints
- [ ] Lava Creek ash
- [ ] left-slip fault
- [ ] limestone
- [ ] marble
- [ ] megabreccia block
- [ ] mud cracks
- [ ] natural bridge
- [ ] Noonday Formation
- [ ] normal fault
- [ ] obsidian
- [ ] pahoehoe lava
- [ ] peridotite
- [ ] petroglyph
- [ ] playa
- [ ] Poleta Formation
- [ ] pothole
- [ ] pumice
- [ ] pupfish
- [ ] rhyolite
- [ ] right-slip fault
- [ ] ripple marks
- [ ] sag pond
- [ ] salt hexagons
- [ ] salt pan
- [ ] sand dune
- [ ] slot canyon
- [ ] steam explosion crater
- [ ] strandline
- [ ] strike-slip fault
- [ ] syncline
- [ ] tephra
- [ ] trace fossil
- [ ] tufa
- [ ] tumulus
- [ ] turtleback
- [ ] upside-down sedimentary rocks
- [ ] vesicle
- [ ] volcanic bomb
- [ ] volcanic crater
- [ ] volcanic dome
- [ ] welded tuff